Analogue Neural VLSI

CHAPMAN & HALL NEURAL COMPUTING SERIES

Series editors: Igor Aleksander, Imperial College, London
Richard Mammone, Rutgers University, New Jersey, USA

Since the beginning of the current revival of interest in Neural Networks, the subject is reaching considerable maturity, while at the same time becoming of interest to people working in an increasing number of disciplines. This series seeks to address some of the specializations that are developing through the contributions of authoritative writers in the field. This series will address both specializations and applications of neural computing techniques to particular areas.

1. **Delay Learning in Artificial Neural Networks**
   Catherine Myers
2. **Analogue Neural VLSI**
   A pulse stream approach
   Alan Murray and Lionel Tarassenko
3. **Neurons and Symbols**
   The stuff that mind is made of
   Igor Aleksander and Helen Morton
4. **Artificial Neural Networks for Speech and Vision**
   Edited by Richard J. Mammone

UNIVERSITY OF STRATHCLYDE

30125 00457785 3

Books are to be returned on or before
the last date below.

20 APR 1994

09 MAR 1995

Analogue Neural VLSI
A pulse stream approach

Alan Murray
Reader in Electrical Engineering
University of Edinburgh
UK

and

Lionel Tarassenko
Department of Engineering Science
University of Oxford
UK

CHAPMAN & HALL
London · Glasgow · New York · Tokyo · Melbourne · Madras

Published by Chapman & Hall, 2–6 Boundary Row, London SE1 8HN, UK

Chapman & Hall, 2–6 Boundary Row, London SE1 8HN, UK

Blackie Academic & Professional, Wester Cleddens Road, Bishopbriggs, Glasgow G64 2NZ, UK

Chapman & Hall Inc., One Penn Plaza, 41st Floor, New York NY10119, USA

Chapman & Hall Japan, Thomson Publishing Japan, Hirakawacho Nemoto Building, 6F, 1-7-11 Hirakawa-cho, Chiyoda-ku, Tokyo 102, Japan

Chapman & Hall Australia, Thomas Nelson Australia, 102 Dodds Street, South Melbourne, Victoria 3205, Australia

Chapman & Hall India, R. Seshadri, 32 Second Main Road, CIT East, Madras 600 035, India

First edition 1994

© 1994 Alan Murray and Lionel Tarassenko

Typeset in Times 10/12pt by Interprint Limited, Malta
Printed in Great Britain by St Edmundsbury Press, Bury St Edmunds, Suffolk

ISBN 0 412 45060 7

Apart from any fair dealing for the purposes of research or private study, or criticism or review, as permitted under the UK Copyright Designs and Patents Act, 1988, this publication may not be reproduced, stored, or transmitted, in any form or by any means, without the prior permission in writing of the publishers, or in the case of reprographic reproduction only in accordance with the terms of the licences issued by the Copyright Licensing Agency in the UK, or in accordance with the terms of licences issued by the appropriate Reproduction Rights Organization outside the UK. Enquiries concerning reproduction outside the terms stated here should be sent to the publishers at the London address printed on this page.
The publisher makes no representation, express or implied, with regard to the accuracy of the information contained in this book and cannot accept any legal responsibility or liability for any errors or omissions that may be made.

A catalogue record for this book is available from the British Library

Library of Congress Cataloging-in-Publication data available

∞ Printed on permanent acid-free text paper, manufactured in accordance with ANSI/NISO Z39.48-1992 and ANSI/NISO Z39.48-1984 (Permanence of Paper).

To Annie and Glynis

Contents

Preface		xi
1 Why build neural networks in analogue VLSI?		**1**
1.1	Introduction	1
1.2	Hopfield memories – the first generation of neural network VLSI	1
1.3	Pattern classification using neural networks	4
	1.3.1 Single-layer networks	5
	1.3.2 Multi-layer perceptrons	8
	1.3.3 Conclusion	10
1.4	Why build neural networks in silicon?	11
1.5	Computational requirement	13
	1.5.1 Digital or analogue?	13
2 Neural VLSI – A review		**15**
2.1	Introduction	15
2.2	MOSFET equations – a crash course	15
2.3	Digital accelerators	19
2.4	Op-amps and resistors – a final look	21
2.5	Subthreshold circuits for neural networks	22
2.6	Analogue/digital combinations	24
2.7	MOS transconductance multiplier	25
2.8	MOSFET analogue multiplier	26
2.9	Imprecise low-area 'multiplier'	27
2.10	Analogue, programmable – Intel Electronically-Trainable Artificial Neural Network (ETANN) chip	27
2.11	Conclusions	28
3 Analogue synaptic weight storage		**29**
3.1	Introduction	29
3.2	Dynamic weight storage	29

3.3	MNOS (Metal Nitride Oxide Silicon) networks	30
3.4	Floating-gate technology	33
3.5	Amorphous silicon (α-Si) synapses	35
	3.5.1 Forming at higher temperatures	36
	3.5.2 Deposition of metal during α-Si growth	36
	3.5.3 Investigation of the forming process	37
	3.5.4 Programming technology	37

4 The pulse stream technique — 38
4.1	Introduction	38
4.2	Pulse encoding of information	39
	4.2.1 Pulse amplitude modulation	41
	4.2.2 Pulse width modulation	41
	4.2.3 Pulse frequency modulation	43
	4.2.4 Phase or delay modulation	43
	4.2.5 Noise, robustness, accuracy and speed	43
4.3	Pulse stream arithmetic – addition and multiplication	44
	4.3.1 Addition of pulse stream signals	44
	4.3.2 Multiplication of pulse stream signals	47
	4.3.3 Interfacing to addition	49
4.4	Pulse stream communication	49
	4.4.1 Asynchronous intercommunication using pulse time information	51
4.5	Conclusions	53

5 Pulse stream case studies — 54
5.1	Overall introduction to case studies	54
	5.1.1 Introduction – Edinburgh SADMANN/EPSILON work	54
5.2	The EPSILON (Edinburgh Pulse-Stream Implementation of a Learning-Oriented Network) chip	55
5.3	Process invariant summation and multiplication – the synapse	55
	5.3.1 The transconductance multiplier	56
	5.3.2 A synapse based on distributed feedback	58
	5.3.3 The feedback operational amplifier	61
	5.3.4 A voltage integrator	61
	5.3.5 The complete system	63
5.4	Pulse frequency modulation neuron	64
	5.4.1 A pulse stream neuron with electrically adjustable gain	66
5.5	Pulse width modulation neuron	67
5.6	Switched-capacitor design	69
	5.6.1 Weight linearity	70

	5.6.2	Weight storage time	70
	5.6.3	Accuracy of computation	71
5.7	Per-pulse computation		71
	5.7.1	Design overview	72
	5.7.2	Input stage	73
	5.7.3	Synapse	73
	5.7.4	Summation neuron	74
	5.7.5	Sigmoid function	75
	5.7.6	Pulse regeneration	75
	5.7.7	SPICE simulation	75
	5.7.8	Results from test chips	76
	5.7.9	Synapse linearity	78
	5.7.10	Input sample and hold	78
	5.7.11	Sigmoid transfer function	81
	5.7.12	Output pulse stream generation	82
	5.7.13	Weight precision	83
	5.7.14	Weight update	84
	5.7.15	Per-Pulse Computation – Summary	84
5.8	EPSILON – The chosen neuron/synapse cells, and results		85
	5.8.1	The EPSILON design	86
	5.8.2	Synapse	87
	5.8.3	Neurons	88
	5.8.4	EPSILON specification	90
	5.8.5	Application – vowel classification	91
5.9	Conclusions		92

6 Application examples — 94
  6.1 Introduction — 94
  6.2 Real-time speech recognition — 94
  6.3 Applications of neural VLSI — 96
  6.4 Applications of neural VLSI – dedicated systems — 96
      6.4.1 Path planning — 98
      6.4.2 Localization — 99
      6.4.3 Obstacle detection/avoidance — 101
      6.4.4 Conclusion — 102
  6.5 Applications of neural VLSI – hardware co-processors — 102
  6.6 Applications of neural VLSI – embedded neural systems — 103
  6.7 Conclusion — 103

7 The future — 104
  7.1 Introduction — 104
  7.2 Hardware learning with multi-layer perceptrons — 105
  7.3 The top-down approach: Virtual Targets — 106

	7.3.1 'Virtual Targets' Method – In an $I{:}J{:}K$ MLP network	107
	7.3.2 Experimental results	108
	7.3.3 Implementation	115
7.4	The bottom-up approach: weight perturbation	116
7.5	Test problem	117
7.6	Weight perturbation for hardware learning	119
7.7	Back-propagation revisited (for the final time?)	121
7.8	Conclusion	124
7.9	Noisy synaptic arithmetic – an analysis	125
	7.9.1 Mathematical predictions	126
	7.9.2 Simulations	127
	7.9.3 Prediction/verification	129
	7.9.4 Generalization ability	129
	7.9.5 Learning trajectory	132
7.10	Noise in training – some conclusions	133
7.11	On-chip learning – conclusion	134

References 136

Index 142

Preface

Biological neural networks have interested scientists for centuries – offering as they do the key to many of mankind's actions, motivations and evolutionary processes. During the 1940s, mathematical models were built of **synthetic** neural networks – computational structures based around a very simple approximation to the human nervous system. They became the subject of grandiose claims, expectations and aspirations. Principally as a result of the work of Minsky and Papert in the late 1960s, however, synthetic neural nets fell into disrepute for some 20 years, during which time only a relatively small number of hardy enthusiasts persevered in what had become an out-of-favour line of research. Around 1980, synthetic neural networks blossomed once more, and the rise in activity between 1980 and 1985 was meteoric. We believe there were fundamentally three reasons for this resurgence. Firstly, the power of conventional computers had advanced to a level at which non-trivial neural structures could be modelled, simulated and proven within a reasonable (by the standards of the day) timescale. Secondly, John Hopfield [1] published a paper on a highly stylised network (the Hopfield network) which appeared to have near-miraculous abilities to capture knowledge via a simple learning scheme. In retrospect, the claims made at the time (mostly by others – not Hopfield) for the network were exaggerated, and the Hopfield network has since become largely a fertile field for theoretical studies. What Hopfield's seminal paper did achieve, however, was to capture the imagination of the physicists (who could see the links between Hopfield nets and physical phenomena), the engineers (who for the first time could understand a neural paper) and the computational scientists (who could simulate neural networks). In this way, Hopfield galvanized a whole new 'neural community' into being. The third piece, which completes the neural renaissance is, we believe, the emergence of technologies (silicon and optical) dense and accurate enough for neural networks to be implemented as hardware.

Which is where we come in. We became seriously interested in synthetic neural networks independently in 1985 (AFM) and 1987 (LT). We came together in 1988 to push forward the pulse stream idea that underlies much of this book. It is fair to say that the work done since then has been performed by a single research group split between two universities, and separated only by geography. As time has progressed, we have each developed individual interests, over and above our shared work in analogue neural VLSI. The collaboration, which emerged spontaneously, and has been rewarding in so many ways, will continue into the foreseeable future. Such natural commonality of interests and views is rare, in a research climate that so often depends upon marriages of convenience to prise support from the hard-pressed funding bodies.

In this preface, it is perhaps worth commenting on the reasons why we set off down the pulse stream path initially. We had both toyed with digital neural ideas for a short while – long enough to realize that digital multipliers were big things, and that the black art that is analogue design was much more intriguing. No 'good analogue process' was available, and our own analogue design expertise was severely limited. Two observations – that biological systems use pulses, and that oscillating circuits (usually a nuisance in VLSI) were not beyond our reach – spawned the initial, admittedly clumsy, pulse stream circuits. Since that stumbling start, many other advantages of pulses as a medium have forced themselves upon us. We are now firmly committed to pulse stream methods as **one of** the most efficient and effective ways of implementing neural VLSI. The emphasis in the preceding sentence is important. There are many workers in this area now, many of whom have produced fine examples of technological, architectural or design cleverness. We make no attempt to tout our own particular work as the best, the cheapest, or the cleverest. Rather we offer it up as a technique that we have found rewarding and powerful.

This brings us to the purpose of this book. Although we are not the first pulse enthusiasts, we were the first to use them in the neural context, and in the way described in these pages. This book pulls together some of our results, circuits and problems, in a form intended to equip the interested reader to develop pulse stream variants appropriate to his or her own application. The book is organized as follows.

In the first three chapters, we attempt to set the analogue neural VLSI context within which the pulse stream work sits. We address the knotty question 'Why analogue VLSI?' in Chapter 1, and then go on to look at the less contentious 'How analogue VLSI?' question in Chapters 2 and 3. Chapters 4 and 5 contain most of the technical information regarding pulse stream, in two complementary forms. While Chapter 4 draws out the generic issues and approaches, Chapter 5 presents a series of case studies, and has been largely written by the designers involved. In this way, we hope to impart a more authoritative and practical feel for how pulse stream works than is possible in a more general treatment. In the concluding chapters we look at two of the areas that are taking up most of our energies now –

applications and hardware learning. In both cases, we have to admit to being in the early stages of work. Nevertheless, we have made some useful advances, and we believe we have some good ideas for future directions. We hope that you will find this book useful and interesting, and if we succeed in motivating one or two with a passing interest in neural VLSI to take their enthusiasm further, we will have achieved what we set out to do.

At the risk of concluding this preface with a section that sounds like an Oscar-winner's speech, we do have to thank a large group of co-workers, encouragers and funding bodies. During the course of the pulse stream work, we have been privileged to work alongside some of the country's brightest and most creative young research engineers – Donald Baxter, Mike Brownlow, Zoe Butler, Stephen Churcher, Alister Hamilton, Jake Reynolds, Tony Smith and Jon Tombs. Thanks, lads and lass. We have received unremitting support from our respective departments – Professors Mike Brady and Peter Denyer have been particularly encouraging, both at the personal and technical level, and Dr Martin Reekie has been a constant source of circuit ideas, healthy scepticism and pragmatism. We have also been fortunate in receiving financial support from both the Science and Engineering Research Council, the CEC (ESPRIT NERVES project), and from Thorn-EMI, British Aerospace, British Telecom, RSRE Malvern and Sharp Laboratories of Europe. Finally, and most importantly, we must thank whatever force caused us to work together in the first place, and our wives and children for putting up with our shuttling up and down between Edinburgh and Oxford. It will continue long into the future.

Alan Murray
Lionel Tarassenko

1

Why build neural networks in analogue VLSI?

1.1 Introduction

The current revival of interest in artificial neural networks can be traced back to two papers published in the early 1980s by John Hopfield from Caltech [1, 2]. The concept of networks of simple threshold elements ('neurons') being used to process input patterns was not new (see *Neurons and Symbols – The stuff that mind is made of* by Aleksander and Morton in this series, for example) but Hopfield introduced the idea of a non-linear *feedback* network such that the inputs to any one neuron were the weighted outputs from every other neuron. For a Hopfield network (Figure 1.1) to have useful computational properties (rather than just exhibit chaotic behaviour), the connection, or 'synaptic', weights between two neurons have to be symmetrical, i.e. $T_{ij} = T_{ji}$, where T_{ij} is the synaptic weight from neuron j to neuron i. In the original, stochastic model [1], each neuron samples its input at random times and changes the value of its output (or not) according to a simple threshold rule:

$$V_i \to +1 \quad \text{if} \quad \sum_j T_{ij} V_j \geq \theta \quad \text{or} \quad V_i \to -1 \quad \text{if} \quad \sum_j T_{ij} V_j < \theta \tag{1.1}$$

where V_i is the state of neuron i and θ is usually set to zero.

1.2 Hopfield memories – the first generation of neural network VLSI

A Hopfield network can be configured to be a *content-addressable memory* if the synaptic weights are generated using the following rule:

$$T_{ij} = \sum_{s=0}^{M-1} V_i^s V_j^s \tag{1.2}$$

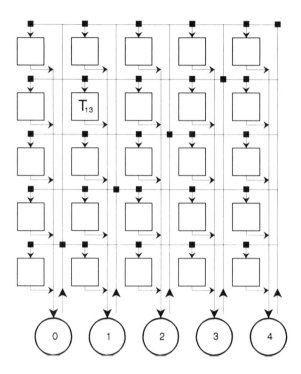

Figure 1.1 *A 5-neuron Hopfield network.*

for M patterns to be stored in the memory. Of course, there must be as many neurons in the network as there are bits in the patterns.[1] On recall, the pattern $\mathbf{v} = \{V_i\}$ is regenerated from its incomplete (or corrupted) version $\hat{\mathbf{v}}$ by computing $\mathbf{T}\hat{\mathbf{v}}$ (where $\mathbf{T}$ is the matrix of synaptic weights), thresholding the result as indicated in equation (1.1), feeding it back to the input and iterating until the network settles into the stable state $\mathbf{v}$. Hopfield summarizes this as follows:

the content-addressable memory ... correctly yields an entire memory from any subpart of sufficient size. [1]

In a second paper [2], a continuous, deterministic model was introduced which retained all the significant behaviours of the original model. The equation of motion for a neuron i in this model is given by:

$$C_i \frac{du_i}{dt} + \frac{u_i}{R_i} = \sum_j T_{ij} V_j \qquad (1.3)$$

---

[1] If the more usual binary coding of 1 and 0 is adopted, the storage rule must be slightly modified, and the number of patterns which can be stored in the memory is halved.

In equation (1.3) u_i is the input potential to neuron i and $V_i = g(u_i)$ where g is a 'sigmoid' function, i.e. a function which is linear for small values of u_i but which saturates at $+1$ for large positive values and -1 for large negative values.

The deterministic model has the same flow properties in its *continuous* space that the stochastic model does in its *discrete* space. In the high-gain case (i.e. for a steep sigmoid), the stable states of the continuous, deterministic system correspond to the stable states of the stochastic system. In a key section of the paper, Hopfield pointed out that equation (1.3) also describes the behaviour of a resistively-connected network of electronic amplifiers, such as the one shown in Figure 1.2, and this comment triggered off a veritable avalanche of attempts at designing Hopfield memories in analogue VLSI. In the circuit, the 'neuron' is represented by an operational amplifier with input resistance ρ_i; the magnitude of the synaptic connection T_{ij} is $1/R_{ij}$, where R_{ij} is the resistor connecting the output of neuron j to the input line i and, finally, C_i is the total input capacitance for neuron i (stray capacitance + amplifier input capacitance). The output voltage from the operational amplifier represents the activity of the model neuron and the currents through the wires, and resistors of a network of several such circuits represent the flow of information through the network. (Note that the possibility that any weight may be inhibitory requires the provision of an additional inverting amplifier and a negative signal wire.)

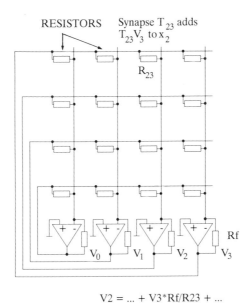

Figure 1.2 *A 4-neuron Hopfield network, implemented as a resistor array.*

Although relatively few of these designs were actually fabricated, they, in effect, were the first generation of neural network VLSI circuits. In one of the very first designs to make it into silicon, networks of operational amplifiers were laid down on a silicon substrate and the synaptic weights between amplifiers were implemented with electron-beam programmable resistive interconnects [3]. The use of these devices, however, was considerably limited by the fact that the networks were non-programmable: the weights were 'locked' in silicon and the devices therefore had fixed functionality.

Since there is no standard technique to fabricate programmable resistors in VLSI[2], the idea put forward by Hopfield of using op-amps and resistors was soon abandoned in favour of MOS transconductance amplifier techniques (see Chapter 2). With these, both the neural state V_j and the synaptic weight T_{ij} are input *voltages* applied to the amplifier which generates an output current proportional to $T_{ij}V_j$.

Despite the proliferation of published designs for analogue VLSI Hopfield networks from 1985 onwards, the fact remains, however, that in 1994 no serious research team is any longer attempting to build Hopfield memories in silicon. There are two main reasons for this.

In the first instance, it was soon realized that the storage capacity of a Hopfield network is extremely poor when compared to a conventional content-addressable memory [4]. If perfect recall is required (i.e. no spurious outputs), then only $n/4 \log_e n$ patterns can be stored [5]. In the case of 64-bit patterns, for example, this corresponds to three or four exemplar patterns only. If a probability of recalling the correct pattern of 95% is acceptable (which is very unlikely in most engineering applications), the number of patterns which can be stored in the network increases to $0.14n$ [6], still a very low number. It is now well established that other types of neural network associative memories are far more efficient, the only advantage of a Hopfield network being the distributed nature of the information storage [7]. The second reason for the drop in interest in Hopfield networks was the re-discovery by Rumelhart *et al.* in 1986 [8] of the *back-propagation algorithm* for the training of multi-layer perceptrons,[3] and this has subsequently proved to be a much more significant development. It has led to the application of feedforward networks to a wide range of problems in pattern recognition or classification over the last few years.

1.3 Pattern classification using neural networks

Pattern classification consists in assigning input data to one of a number of M classes. The usual method is to have M output units and use a

---

[2] Switched-capacitor implementations are not feasible either, as this would require as many Voltage Controlled Oscillators (VCOs) as there are synapses.

[3] Re-discovery rather than invention because it had first been proposed by Werbos in 1974 but not exploited by him or anyone else at the time.

1-out-of-M coding for the output vector: during training, the unit corresponding to class i is given a target value of 1, all other units being assigned a value of 0; during testing with the trained network, the class of the input pattern is determined by identifying the output unit with the maximum value. *Training*, or learning, proceeds by adjusting the network weights to obtain an appropriate mapping between the input patterns in the training set and the corresponding output vectors. *Generalization* can be said to take place if the network assigns the right class to an input pattern which was **not** in the training set. Of course, this definition on its own is not enough as it does not take into consideration the complexity of the task nor the statistical distribution of the patterns both in the training and the test data sets. For good generalization, the training set should contain a number of patterns P which is at least several times larger than the network's capacity, W/M, where W is the number of weights in the network [9].

1.3.1 Single-layer networks

The simplest decision unit is the *perceptron* (Figure 1.3), a threshold logic unit which assigns an input pattern to class k or not to class k. (An M-class problem requires M perceptrons operating in parallel on the same input patterns.) Each element of the input pattern is multiplied by a synaptic weight T_{kj} and the resulting sum is passed through a hard-limiting non-linearity. Thus, if we use the same notation as for the Hopfield networks

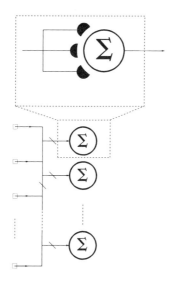

NI INPUTS   NC OUTPUT

Figure 1.3 *NI:NC single layer perceptron network.*

described at the beginning of this chapter, the output V_k of perceptron k will be given by:

$$V_k = f_h\left(\sum_j T_{kj} V_j\right) \tag{1.4}$$

where V_j is the jth element of the input pattern and f_h is the hard-limiting function such that:

$V_k = 1$ if the input pattern belongs to class k

$V_k = 0$ if the input pattern does not belong to class k

At the start of training, the synaptic weights to each perceptron in the network are initialised with small random values. The *perceptron learning rule* [10] is an error-correction procedure which requires the values of the T_{kj} weights to be adapted according to the following equation:

$$\Delta T_{kj} = \eta V_j (D_k - V_k) \tag{1.5}$$

where η is a parameter called the learning rate (usually $0.01 < \eta < 0.1$) and D_k is the desired output, i.e. 1 for class k, 0 otherwise. Since V_k itself can only be 0 or 1, the weight update ΔT_{kj} can only be $\pm \eta V_j$ or 0. During training, the input patterns are presented to the network in random order and the above equation is used to adjust the weights for each pattern in turn until all the patterns are correctly classified, *if such a solution exists*. This is the case if the input patterns are *linearly separable* into the M required classes, that is if it is possible to separate the j-dimensional input patterns with $(k-1)$-dimensional hyperplanes.[4] The proof of this, the perceptron convergence theorem, was first given by Minsky and Papert [11]. They showed, however, that the perceptron learning rule fails to find a solution to problems for which the input patterns are **not** linearly separable; in those cases, the weights continue to oscillate without ever converging. The best known example of such a problem is the Exclusive-OR problem where, of course, the four possible 2D input patterns cannot be separated into class k or not k simply by drawing straight lines through the square which has the (0, 1), (1, 0), (0, 1) and (1, 1) input patterns as its corners.

It is by now well documented that Minsky and Papert's exposition of this limitation caused neural network research to stall for more than a decade. It is often forgotten, however, that the Least-Mean Square (LMS) learning algorithm[5] does **not** lead to an unreasonable weight solution if the patterns

---

[4] The term linear separability arises from the two-dimensional case: for 2D input patterns, the criterion is met if lines can be drawn in input space to separate the training patterns into their M classes.

[5] Proposed independently by Widrow and Hoff (1959) at almost the same time as Rosenblatt's perceptron learning rule.

are not linearly separable. The LMS learning rule does not attempt directly to separate the different classes, rather it seeks to minimize the mean squared output error $\frac{1}{2}\Sigma(D_k - V_k)^2$ according to the following equation:

$$\Delta T_{kj} = \frac{\eta}{|V_j|^2} V_j (D_k - A_k) \qquad (1.6)$$

where $A_k = \Sigma_j T_{kj} V_j$ and $V_k = f_h(A_k)$. Since A_k is unbounded, the weight update is normalized by the $|V_j|^2$ factor.

Although the LMS rule may fail to separate training patterns which are linearly separable, it is our opinion that it should be used in preference to the perceptron learning rule as it gives an acceptable solution **on all problems**. It is often the case, especially for problems with high input dimensionality, that a single-layer network trained with the LMS rule gives a classification error rate on test patterns which is at least as good as, if not better than, multi-layer networks for which the ratio W/M is bound to be much higher for the same problem. Difficult pattern recognition problems, however, require complex decision surfaces and multi-layer perceptrons (see Section 1.3.2) will be required to obtain these.

An alternative for problems with input data of low dimensionality is to construct, with a non-linear mapping, a representation in a space of higher dimension. This representation can then be used as the input to a single-layer network trained with the LMS algorithm. This is exactly the approach which is adopted with Radial Basis Function (RBF) classifiers (Figure 1.4).

These classifiers create complex decision regions from kernel function nodes which form overlapping receptive fields. These nodes are the hidden units of the RBF classifier; they compute radially symmetric functions (usually Gaussians) which give a maximum when the input pattern is near in Euclidean distance to the prototype pattern stored in that node's weights. Training usually consists in optimizing the location in input space of these prototype patterns by using a clustering procedure (***unsupervised*** learning) such as the adaptive K-means algorithm [12][6] or Kohonen's feature map algorithm [13].

RBF classifiers typically give results which are comparable to those obtained with multi-layer perceptrons. The rest of this chapter, and indeed of the book, concentrates on the latter as our design effort has, so far, been focused on linear summation and sigmoid circuits. The VLSI implementation of RBF networks requires circuits for the computation of Euclidean distance and of Gaussians; the interested reader is referred to the work of Platt, for example, for details of such circuits [14].

---

[6] More complicated procedures which seek to optimize both the location of the centres and their covariance matrices can also be used; the only advantage, in general, is the fact that the same performance can be obtained with slightly fewer centres.

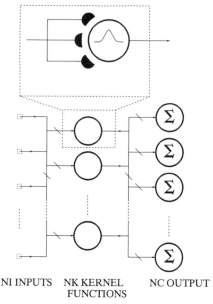

Figure 1.4 *NI:NK:NC radial basis function network.*

1.3.2 Multi-layer perceptrons

Multi-layer perceptrons (MLPs – Figure 1.5) are *feed-forward networks* with one or more layers of hidden units between input and output. It has recently been shown that a network with a single layer of hidden units is sufficient to approximate to any desired accuracy any non-linear input/output function, but this is a 'weak theorem' in the sense that it says nothing about the **number** of units required in the hidden layer to approximate a given function within a certain error.

The transfer function of the neurons in the multi-layer perceptron is a sigmoid function. Different such functions can be used, the only requirement being that they should be monotonic and differentiable (so that the chain rule can be applied and errors back-propagated from output to input). The most popular is the logistic sigmoid function, for which:

$$V_k = \frac{1}{1+e^{-x_k}} \quad \text{where} \quad x_k = \sum_j T_{kj} V_j \qquad (1.7)$$

where, again, V_k is the output of neuron k and T_{kj} is a synaptic weight from neuron j in the previous layer.

As with the simple LMS rule, adaptation of the weights is achieved by minimizing the mean square output error defined as:

$$E = \frac{1}{2} \sum_P \sum_{k=1}^{M} |V_k - D_k|^2 \qquad (1.8)$$

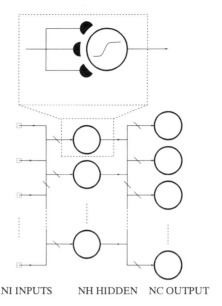

Figure 1.5 *NI:NH:NC multilayer perceptron network with a single hidden layer.*

where, as before, D_k is the desired output for neuron k, M is the number of neurons in the output layer and P is the number of training patterns. The error E is minimized, in most cases by using gradient descent:

$$\Delta T_{kj} = -\eta \frac{\partial E}{\partial T_{kj}} \qquad (1.9)$$

Back-propagation of errors through the network is an exact analytical procedure for evaluating the first derivative of the error with respect to the weights:

$$\frac{\partial E}{\partial T_{kj}} = \frac{\partial E}{\partial x_k} \frac{\partial x_k}{\partial T_{kj}} = \delta_k V_j \qquad (1.10)$$

where $\delta_k = \partial E/\partial x_k$, with $x_k = \Sigma T_{kj} V_j$ and $V_k = f(x_k)$. If the activation function f is a logistic sigmoid, we can write for an output neuron k:

$$\delta_k = \frac{\partial E}{\partial V_k} \frac{\partial V_k}{\partial x_k} = (V_k - D_k) V_k (1 - V_k) \qquad (1.11)$$

where D_k is the desired output and V_k the actual output. If the neuron is an internal neuron j from the hidden layer, then

$$\delta_j = V_j (1 - V_j) \sum_k \delta_k T_{kj} \qquad (1.12)$$

where k is over *all neurons in the layer above the hidden layer*, i.e. the output layer for networks with a single hidden layer. Knowledge of all the δs in the next layer is therefore required to compute the weight update equation for a hidden layer neuron.

Training continues until E has decreased below a given threshold. It can be difficult to decide on the value of this threshold as continued training would in most cases further reduce E but also result in the decision space being adjusted to fit the noise rather than just the data. Such 'over-training' is a common cause of poor generalization with MLPs and the only satisfactory solution to this problem is to divide the data set of input patterns into a training set, which is used for weight adaptation, a cross-validation set for deciding when to stop training, and a test set to evaluate the generalization performance of the trained network.

1.3.3 Conclusion

This section has focused on feedforward networks with a single layer of hidden units. This emphasis has been deliberate. The claims which Hopfield made in his original paper about the 'spontaneous emergence of computational properties' of feedback networks have since been toned down, if not abandoned. In contrast, it is now generally accepted that multi-layer feedforward neural networks are capable of obtaining good solutions to computationally hard problems. Although multi-layer perceptrons and Radial Basis Function classifiers are not the only types of neural network classifiers, they have proved the most popular in the current phase of activity (see the *Proceedings of the IEE International Conferences on Artificial Neural Networks*, for example). As already mentioned in the section on RBF classifiers, there are also a number of neural network **clustering** algorithms which are trained in unsupervised mode. The best-known of these are Kohonen's self-organizing feature map (Figure 1.6) and Grossberg's Adaptive Resonance Theory (ART) networks (Figure 1.7).

Kohonen's algorithm for training the feature map is a clustering algorithm with the additional concept of a neighbourhood function which gradually shrinks over time. As a result, input patterns which are close to each other (in Euclidean distance) activate units on the feature map which are topologically close to each other. The relationship between distances in input space and distances on the 2D feature map obviously cannot be preserved but topological organization of the map is obtained. Grossberg's ART-1 training algorithm is a form of sequential leader clustering. The first input pattern is taken to be the prototype (or exemplar) pattern for the first cluster. The next input pattern is assigned to the same cluster if its distance to the first prototype pattern is within a given threshold; if not, it becomes the prototype pattern for a new cluster. This process continues for all inputs submitted to the network. The number of clusters created during training is

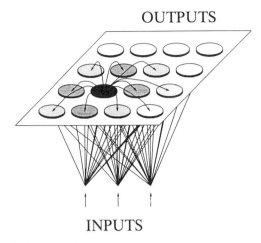

Figure 1.6 *Kohonen's self-organizing feature map architecture (schematic).*

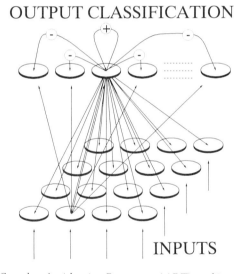

Figure 1.7 *Grossberg's Adaptive Resonance (ART) architecture (schematic).*

obviously very dependent on the value chosen for the threshold. For more details on these and other neural network architectures, the reader is invited to consult the review article written by Lippmann [15].

1.4 Why build neural networks in silicon?

The re-emergence of artificial neural networks can be seen partly as the rediscovery of pattern recognition techniques [16] prematurely abandoned at the end of the 1960s in favour of symbolic AI. There is now a growing realization that not all problems requiring 'computer intelligence' lend themselves naturally to the high-level descriptions and the formal reasoning of symbolic AI. This of itself, however, does not provide any kind of justification for building neural networks in silicon as opposed to running the algorithms on a general-purpose computer. On the other hand, it can also be said that one of the reasons why the current phase of research in artificial neural networks is lasting longer than the previous cycle of interest in pattern recognition has been the advances since then in computer *hardware* which have made it possible for the solution to real-world problems to be tackled and solved in reasonable time.

There are still problems in vision and robotics, for example, which cannot be solved in **real time**, a situation likely to continue until suitable parallel architectures have been developed. One of the motivations, from the earliest days, for research into artificial neural networks has been the fact that they can be mapped directly into massively parallel hardware which can then provide real time solutions for the type of problems requiring very high computational rates. Of perhaps equal importance is the fact that the neural paradigm is highly fault-tolerant, since damage to a few links (synaptic weights) is not catastrophic. Some have argued [17] that this will be of increasing importance in the manufacture of integrated circuits as feature size continues to shrink and the transition is gradually made from Very Large Scale Integration (VLSI) to Ultra Large Scale Integration (ULSI).

These, then, are some of the motivations which have been behind our research into VLSI neural networks over the past five years. Our medium-term goal has been to build simple real-time hardware systems demonstrating the capabilities of neural VLSI and we report on this work in Chapter 6. A longer term goal is to address the question 'how can we perform neural-like computation in silicon which best exploits the characteristics of the available material?' This approach requires the abandonment of current neural network algorithms[7] and a focusing instead on strategies which exploit the characteristics of transistor physics directly 'to compute'. This, in essence, is the approach adopted by Carver Mead and described in his recent book on *Analog VLSI and Neural Systems*. It represents a radical departure from his influential work on digital VLSI techniques in the early 1980s – see, for example [18]. Mead has deliberately sought inspiration from neurobiology and indeed all the designs which have so far been fabricated have attempted to mimic the function of a biological 'transducer', such as the retina or cochlea. Whether this approach can lead to the evolution (and the word is used advisedly)

---

[7] Which, after all, run on conventional computers.

of a new computational style or even paradigm remains to be seen. For our part, we have adopted a more gradualist approach: in the first instance, we have developed basic building blocks for neural computation using the *pulse-stream* technology described in this book; we have then assembled these building blocks to build the real-time demonstrators of Chapter 6, and finally we are about to re-evaluate them in the context of 'hardware learning' (see the final chapter in this book). We believe that the building of neural learning machines in silicon, capable of self-modification, is an important goal in the development of new computational techniques for parallel hardware.

1.5 Computational requirement

The computational requirement for massively parallel neural network hardware leads to conceptually very simple building blocks:

1. Presynaptic neural inputs V_j must be multiplied by synaptic weights T_{kj} and accumulated to produce neural activities $x_k = \Sigma T_{kj} V_j$.
2. These post-synaptic activities must subsequently be passed through a non-linear activation function $f(\ )$ to produce an output neural state $V_k = f(x_k)$.

For example, a 100-50-100 multi-layer perceptron network requires 10^4 multiply/add operations, and 150 non-linearity operations for each forward pass. Clearly, the multiply/add (synaptic) operation represents the biggest computational load, and the efficiency and hardware 'cost' of the implementation of this building block are the key issues in the hardware design of neural networks.

1.5.1 Digital or analogue?

The main advantage of a digital implementation is its predictable accuracy. However, digital multipliers occupy large areas in silicon and consume appreciable amounts of power. As a result, the application of digital hardware is likely, for the foreseeable future, to be restricted to **hardware acceleration** of software simulations. If artificial neural networks are to be implemented on a **fully parallel** architecture for the reasons given in the previous section, then **analogue** VLSI is required. Analogue multipliers can be built with as few as three transistors (see next chapter), and it is therefore possible to have several tens of thousands of multipliers working in parallel on one chip. (It must be remembered, however, that very high computational rates count for very little if one is limited by the bandwidth of input and output data transfers.) Analogue computation does sacrifice the known accuracy of digital circuitry, but the nature of neural computation should ensure that this is not necessarily as much of a disadvantage as it may appear, and we will return to this issue in the last chapter. In the medium term, we are prepared to trade the precision of a single floating-point digital multiplier for the massive parallelism possible with large numbers of analogue multipliers. In the longer term, analogue operation offers the only hope of exploiting device physics to the full in any neurally-inspired attempts to evolve new computational styles.

2

Neural VLSI – a review

2.1 Introduction

This is not intended to be a 'textbook', and this chapter does not, therefore, purport to give an exhaustive review of neural VLSI. We have chosen, rather, to provide a 'good example' of each of the four major genres of neural devices – digital, fixed function analogue, hybrid analogue/digital and purely analogue, programmable. These are introduced as their respective implementation technologies are described.

Since standard VLSI technology is predominantly based on Metal Oxide Silicon Field Effect Transistors (MOSFETs), it is worth reviewing their behaviour, or at least the aspects of their behaviour which are relevant to neural VLSI. The reader should be aware by now that this means concentrating on *analogue* VLSI and studying the characteristics of MOS devices over the full range of input/control parameters.[1] This section, therefore, is aimed at providing readers without any MOSFET VLSI experience with the bare minimum of transistor theory to facilitate understanding of the more detailed sections that follow in this and later chapters. The treatment thus makes no attempt to be either comprehensive or physically justified, and readers **with** MOS experience can skip this description completely. Those readers, on the other hand, who would like more details are referred to a suitable book on analogue VLSI design such as [19].

2.2 MOSFET equations – a crash course

MOSFETs come in two basic flavours – N-type and P-type (N for Negative, P for Positive). Current in N-types is carried by negative electrons, while current in P-types is carried by positive 'holes'. An N-type device is formed in and on top of a P-type substrate, and *vice versa*. The MOSFET is a four terminal device – the terminals are the Gate, Source, Drain and Substrate.

---

[1] Which may cover several orders of magnitude – see the later section on **subthreshold circuits** in this chapter.

For our purposes, it will be enough to describe its functionality qualitatively, and follow this with a few equations that encapsulate the important aspects of MOSFET behaviour.

Taking the N-type MOSFET first (Figure 2.1) – current flows from drain-source if a voltage is applied across the drain-source. Conventionally, we refer to the lower voltage terminal of an N-type MOSFET as the source, and the higher as the drain, although they are physically no different. No DC current flows into the gate, and only very small currents flow into the substrate. If the voltage between the gate and the source is below a value called the transistor **threshold**, the current flowing from source to drain is also very small (although not zero). In this mode, the transistor is said to be OFF. If the gate-source voltage is above the threshold, the drain-source current is a function of that voltage – getting bigger as the gate-source voltage gets bigger. It is neither necessary nor desirable that we become embroiled in device physics here, but it is interesting to note that this behaviour depends critically on the positive gate-source voltage attracting negative carriers from the P-type substrate to the surface, where they can act as the carriers of the drain-source current. We might therefore expect a P-type device to be something of a mirror-image, with respect to its operational characteristics.

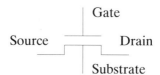

Figure 2.1. *N-channel MOSFET.*

For a P-type device, the source is the more positive and the drain the more negative terminal. The gate must be more than a transistor threshold **below** the source for the MOSFET to be on (conducting, source-drain). We can thus more or less alternate between thinking about N- and P-type devices by changing polarities everywhere, and calling the terminals the appropriate names. This will become clearer when we look at the device equations.

For now, let us look at an N-type MOSFET with its source and bulk (substrate) connected together, $V_{BS}=0$. For such a device, the all-important threshold voltage is given by:

$$V_T = V_{FB} + 2\Psi_B + \frac{\sqrt{2q\varepsilon_{Si}N_A 2\Psi_B}}{C_{ox}} \pm \frac{Q_{ss}}{C_{ox}} \qquad (2.1)$$

where Ψ_B is the voltage difference between the intrinsic Fermi level and that of the doped silicon, and V_{FB} is the so-called 'flat-band' voltage – both characteristics of the silicon substrate itself [20].

This is a messy equation, but it need not concern us much. All that we need to note from it here is that it contains a large number of process-determined factors. N_A, for example, is the density of acceptor impurities in the silicon substrate – a quantity determined by exactly how, and for how long, the impurity implant process is performed. Typically, transistor thresholds are around 1 V, in a process designed for 5 V operation. Unfortunately, even turned-ON MOSFETs do not obey a single equation with respect to drain-source current. When the gate-source voltage is large, and the drain-source voltage relatively small, the MOSFET is said to be in its *linear* region, and obeys the equation:

$$I_{DS} = \frac{\mu C_{ox} W}{L} \left[(V_{GS} - V_T) V_{DS} - \frac{V_{DS}^2}{2} \right] \quad (2.2)$$

When the drain-source voltage is raised (or the gate-source voltage lowered) such that $V_{DS} > V_{GS} - V_T$, the transistor becomes **saturated**, and is governed by:

$$I_{DS}(sat) = \frac{\mu C_{ox} W}{2L} (V_{GS} - V_T)^2, \quad (2.3)$$

where

C_{ox} = oxide capacitance/area – determined by the fabrication process
μ = carrier mobility – determined by the fabrication process
W = Transistor gate Width – determined by the MOS designer
L = Transistor gate Length (Source→Drain) – determined by the MOS designer.

Both equations (2.2) and (2.3) become invalid if the transistor is OFF ($V_{GS} < V_T$). We will return to what happens then later. Both equations contain a term that can be adjusted by changing the transistor length or width. This is the VLSI designer's primary influence on what goes on. To see what the two equations mean, it is worth looking at graphs showing the drain-source current for different drain-source and gate-source voltages, as shown in Figure 2.2. For constant drain-source voltage, nothing much

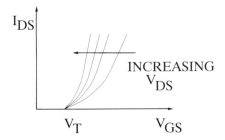

Figure 2.2. *Drain-source current as a function of gate-source voltage for an n-channel MOSFET for varying drain-source voltages.*

happens until V_{GS} exceeds V_T. Above this point, the current through the MOSFET rises non-linearly, and quite rapidly, with increasing V_{GS}. As V_{GS} rises *just* above V_T, the transistor is in the saturated condition, and is obeying equation (2.3). As V_{GS} rises above $V_{DS} + V_T$, the transistor is in the linear region, and is governed by equation (2.2). It is more common, for reasons that are not immediately obvious, to plot I_{DS} against V_{DS}, for different values of V_{GS}. This is shown in Figure 2.3. You will notice that the current rises initially, as V_{DS} is increased from 0 V, almost linearly. This is the term $(V_{GS} - V_T)V_{DS}$ at work, giving the 'linear' region its name. In this region, the MOSFET is behaving essentially as a resistor, with a resistance value given by $[\mu C_{ox} W/L \times (V_{GS} - V_T)]^{-1}$. This will be seen to be a useful property later in this chapter. As V_{DS} increases further and the MOSFET enters the saturation regime, equation (2.3) suggests that the curve should flatten. As you can see, it is almost flat above $V_{DS} > V_{GS} - V_T$ – the slight slope is due to the conducting channel length being modified as V_{DS} rises. This is a second-order effect, albeit an important one. We do not need to understand it here, but we must be aware of it, as it determines how well MOSFETs behave, particularly when used as current sinks, sources, or mirrors. In the MOSFET equations, it is taken account of by multiplying both (2.2) and (2.3) by $(1 + \lambda V_{DS})$, where λ is a small, empirically-determined parameter.

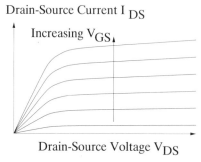

Figure 2.3. *Drain-source as a function of drain-source for an n-channel MOSFET for varying gate-source voltages.*

Modelling transistor behaviour is fraught with this sort of difficulty. There are many other second order effects, whose physical explanations are not all well established. Much of the time, they can be ignored. In designing precision analogue circuits, they must be accounted for via simulation, and the designer needs to make himself aware of the models underlying his simulations – they will be baroque versions of (2.2) and (2.3), in the case of MOSFETs. The only other second-order effect that need concern us here (although they are all important – see [20] or [19] for a comprehensive treatment) is the **body effect**. So far we have glossed over the role of the fourth MOSFET terminal – the **bulk** or **substrate** terminal. Equations (2.1–2.3) all assume that an N-type MOSFET has its bulk and source terminals connected together, probably to 0 V, while a P-type will have its bulk and source connected together, often to 5 V. When V_s and V_B are **not** the same, some more complicated physics results, and the transistor threshold changes. It becomes

$$V_T \rightarrow V_T + \gamma[\sqrt{|2\Psi_B| + |V_{BS}|} - \sqrt{|2\Psi_B|}] \qquad (2.4)$$

In other words, if the N-type source voltage rises above the substrate for any reason, the transistor becomes harder to turn ON, as its threshold voltage V_T is driven up. This can be a serious nuisance in analogue circuits, and a more minor nuisance in digital. In P-types, all of this is upside-down again, and allowing the source voltage to drop below the substrate makes the transistor become harder to turn ON. Before we leave this section, it is worth noting that allowing the source to drop significantly below the substrate in an N-type (above in a P-type) is a very bad idea, as it does much more than simply alter the thresholds. In this situation, parasitic diodes in the silicon which are normally safely reverse biassed, and not passing current, become forward-biassed, and pass large currents. The result is generally destruction of the chip.

The only other effect of importance not discussed above is what happens when the transistor is turned OFF. We have hinted that small but important currents flow. To avoid this section becoming a major exercise in transistor theory, we will defer discussion of this nicety to the discussion of Carver Mead's work, where subthreshold conduction is actually used constructively.

2.3 Digital accelerators

Before plunging into the analogue domain, it is worth reviewing what is currently the state of the art in **digital** neural VLSI. It is clearly possible to implement $\Sigma T_{ij}V_j$, and an appropriate activation function, to arbitrary precision, using digital techniques. The hardware platform can be a conventional Von Neumann, general purpose computer, a 'supercomputer', a general-purpose hardware accelerator such as a fast Digital Signal Processing (DSP) chip, or a custom neural device. For most applications, a general-purpose simulator, implemented on general-purpose hardware, is both appropriate and convenient. We have already explored the question of the need for hardware in Chapter 1.

In this short sub-section, we examine what we consider to be the best custom digital neural VLSI component – the Adaptive Solutions CNAPS device.

The CNAPS (Connected Network of Adaptive Processors) chip is based around the 'Processor Node' shown in Figure 2.4. The node comprises

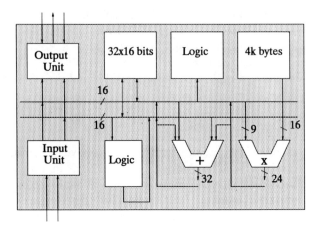

Figure 2.4. *Adaptive Solutions' CNAPS (Connected Array of Adaptive Processors) processor node.*

a single (9-bit × 16-bit) multiplier (to form $T_{ij}V_j$), an adder to perform the summation and 4 bytes of local RAM, along with some 'glue' logic – input/output control and an arithmetic shifter, etc. Each CNAPS comprises 64 Processor Nodes, clocked at 25 MHz. The device is configurable via microcode to define both the network architecture of a CNAPS-based system, and the arithmetic precision of the neural arithmetic itself. In this way, arithmetic precision can be traded off against raw computational speed. Hammerstrom [21] gives the performance figures based on several different combinations, as CPS (Connections Per Second) and also as total number of connections per chip:

At 25 MHz – all PNs busy

12.8 billion CPS for 1 × 1-bit multiplication

1.6 billion CPS for 8 × 8- or 8 × 16-bit multiplication

0.8 billion CPS for 16 × 16-bit multiplication

Total Connections/Chip

2M 1-bit weights

256k 8-bit weights

128k 16-bit weights

16 × 16-bit multiplication is achieved in two cycles of the 9 × 16-bit multiplier. The chip is not huge – 26 × 26 mm, and comprises some 14 M transistors. In all, CNAPS is an impressive device, offering very fast

performance in a 'safe' digital form. While we do not advocate the use of analogue techniques for reasons of speed alone, CNAPS forms a useful benchmark against which speed comparisons should be made.

It is worth noting in passing, however, that Hitachi announced a WSI (Wafer Scale Integration) product that implements 114 digital neurons, each with 64 16-bit weights on a wafer. A multi-wafer system looks like an old-fashioned juke-box, and implements 1152 such neurons. The whole system is implemented as 3-layer metal, 0.5 μm CMOS on 5″ wafers, and consumes some 50 Watts. The economics of such a system are scarifying, as is the massive extra redundancy that WSI implies (it is simply not enough to expect the neural paradigm to take care of circuit imperfections – it will not). The example illustrates, however, what can be achieved when the very best technology is brought to bear upon a problem without regard for cost.

2.4 Op-amps and resistors – a final look

Once a (crystalline or amorphous silicon) resistor is fabricated within a CMOS process its value is fixed, unless non-standard technology[2] or switching circuitry is used to allow its value to be changed electronically. Without such steps being taken, the neural network's functionality is fixed and it was made clear in Chapter 1 that this lack of programmability was the main reason why the op-amps and resistors approach first suggested by Hopfield had been abandoned. In fact, there are specialised applications in which the weight set is known *a priori* and will not be changed. An example of such an application is the front-end circuitry used for feature extraction in the Optical Character Recognition neural chips developed by Bell Labs [22]. The designers here knew from prior software simulation which features to extract (for example corners, line endpoints, etc.) and were therefore able to design a chip with 32,000 reconfigurable *binary synapses*, partitioned in 256 neurons with 128 binary inputs each. Having found that 3 to 4 bits were needed to encode the weights, they then combined the *linear* (i.e. prior to hard-limiting) outputs of several neurons with power-of-two coefficients to provide up to 4 bits of analogue depth.

Limited programmability *can* be introduced into the op-amp and resistors approach if one is prepared to pay the price for it. It is clearly possible to fabricate an *array* of resistors, with a range of values, at each synapse location, switching electronically. There is, however, a very large hardware overhead, and the only significant example of this technique uses a tri-flop cell to allow values ($+1$, 0 or -1) to be programmed for the set of T_{ij}s. Incidentally, both the Bell Labs and Caltech groups involved in this piece of work have subsequently abandoned the approach in favour of fixed-function network integration. The additional complexity and area involved

---

[2] We will return to this issue in Chapter 3.

in including the digital switching circuitry (not forgetting the associated digital memory) make this approach a combination of the worst of digital and analogue characteristics.

2.5 Subthreshold circuits for neural networks

The first foray into neural silicon by Carver Mead's group at Caltech used the op-amp/resistor concept for a Hopfield memory in which each T_{ij} weight (see above) could be set to one of 3 values ($+1$, 0 or -1) at the cost of 41 transistors per synapse. Since this early work, Mead has evolved what amounts to a complete subculture of analogue neural CMOS design methods, circuits and devices based on subthreshold (weak inversion) MOSFET operation [23], although it should be remembered that subthreshold circuits are far from new [24]. It is impossible in a review of this nature to do more than give a flavour of the type of work done. As discussed in the section on MOSFET equations, digital designers tend to view a MOSFET as a device whose drain-source current I_{DS} is zero for $V_{GS} \leqslant V_T$. Analogue designers (and DRAM designers) have long known this to be far from true. A 'turned off' MOSFET continues to pass current from drain to source, albeit of the order of nA ($Amps \times 10^{-9}$). To the creative analogue designer, this is a fascinating feature to be utilized – while to the DRAM designer, who wishes to use the MOSFET as a switch with zero ON resistance and infinite OFF resistance, it is a serious problem. In fact, in this operating regime, drain-source current depends **exponentially** on gate-source voltage:

$$I_{DS} \alpha\, e^{KV_{GS}} \tag{2.5}$$

where K depends on the detailed theory of weak inversion operation. This relationship allows circuits developed for bipolar transistors (see below), whose input-output relationship is similar, to be adapted directly for MOS usage, and the extremely small currents involved reduce the power consumption in subthreshold devices to very low levels. This fact has been used for some time [24] to design circuits with low power consumption. A price is paid, however, in terms of noise immunity. Concerns regarding the noise immunity of such circuits may be answered, in principle at least, by the robustness implied by the massive parallelism in a neural network. Certainly, the biological nervous system copes with noise. Some attempts have been made to analyse the effects of noise in neural networks [25–27], and preliminary studies indicate that noise may indeed be at least tolerable, and perhaps helpful (see Chapter 7) in neural computation. Mead takes the neurobiological analogue one step further, and points out that the dependence of nerve membrane conductance on the potential across the membrane is also exponential [23]. We are not entirely convinced that this admittedly pleasing coincidence is in itself particularly valuable. However, we are in sympathy with Mead's philosophy that MOSFET non-linearities are a feature to be celebrated and utilized, rather than lamented.

Mead's book [23] is a veritable treasure-trove of wisdom, wit and clever circuits. It would be foolish even to attempt to capture any more than the flavour of its contents. Purely as an example of the circuits used by Mead in his work, Figure 2.5 shows the *Gilbert Multiplier* [28], well-known in bipolar design as an efficient, four-quadrant multiplier that produces an output current I_{out} that is proportional to $(V_1 - V_2) \times (V_3 - V_4)$ over a limited operating regime. Furthermore, the product saturates elegantly (via a *tanh* function) as it tries to exceed the bias current I_b. Mead's group has designed many special purpose chips using this and similar circuits and techniques building on biological exemplars – even to the extent of designing a silicon slug [29]. The contrast between Mead's approach and that of (say) Bell Labs [22] illustrates the diversity of opinion that exists in the neural research community over the importance (or otherwise) of the biological exemplar. For some, it is merely the inspiration behind the neural paradigm, and sometimes embarrassing, while others regard it as a splendid working example, that should be adhered to in detail, as well as in general terms. We adopt a compromise position – always aware of the biological exemplar, but not bound by stricture to adhere to its methods.

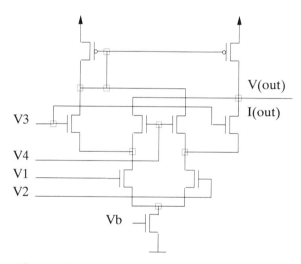

Figure 2.5. *Gilbert multiplier circuit using MOSFETs in their subthreshold (weak inversion) regime.*

Before leaving Mead's work, it is worth noting that his apparently esoteric circuits have created one of the first few genuine neural products – perhaps even the first neural VLSI product. The problem is deceptively simple – to read and classify the stylised characters on a cheque that are used to indicate the bank account number. The number of classes is small (10 numbers) and the text font well-defined. However, cheques get dirty and are scribbled upon. Furthermore, banks often print images **behind** the cheque printing, thus confusing the issue further. Mead was almost uniquely well-placed to tackle this problem, with his experience in producing direct optical-interface devices such as silicon retinas [30], coupled with his wealth of circuits for fixed-weight neural functions. The task is such that the training data are presented once only (to encode the classes to be recognized) and the weights can then be encoded permanently on chip. Subsequent adaptation is neither necessary nor desirable. Figure 2.6 illustrates the system schematically. One device indicates when the image on the optical sensor array is centred, and the other classifies it. The output class is then latched digitally for use by the remainder of the system.

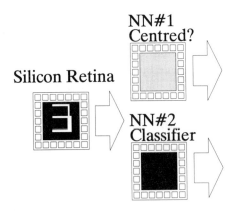

Figure 2.6. *Synaptics' cheque-reader chipset (schematic). A good example of a fixed-function, carefully engineered neural hardware solution to a real problem.*

This is an example of a delightful, and not at all accidental symbiosis between problem and solution, and is an eloquent demonstration of exactly what form of problems will yield to fixed-weight solutions. The problem may, in fact, be much 'harder' than the cheque-reader, but the constraint is the same. Once defined, neither the problem, nor the desired class definitions, must change, and the market must be large enough to justify a totally customised component. The directness of the analogue (sensor) input is purely a result of Mead's own wide capabilities, and once again a perfect example of one of the biggest merits of analogue techniques.

2.6 Analogue/digital combinations

The best example of a successful combination of digital and analogue techniques is that of ANNA – the Analogue Neural Network Accelerator –

from Bell Labs [31]. ANNA is a 4096-synapse, 8-neuron device, using analogue arithmetic under control of digital neural state signals. While it is not a pulsed chip, it is interesting to note that this describes the EPSILON chip, whose description concludes Chapter 5, fairly accurately. Weights are stored as analogue, dynamic voltages, refreshed cyclically from off-chip digital RAM, via on-chip digital-to-analogue converters. They are relatively coarsely quantized (6-bits/weight), and states are only 3-bit words. This is necessary to reduce the size of the arithmetic element, the Mutiplying Digital-to-Analogue Converter (MDAC). The chip occupies some 4.5×7 mm in $0.9\,\mu$m CMOS. The level of integration is thus broadly similar to our own EPSILON device, although ANNA offers a much faster weight-refresh cycle via a clever current-mode refresh circuit. EPSILON has, however, no overt quantization errors. It is too early to say whether EPSILON outperforms ANNA, and as ever the definition of 'outperform' must be considered carefully.

Bell Labs' primary application for ANNA is as a slice of an Optical Character Recognition (OCR) network. The network architecture is highly stylised – consisting of five layers of processing, with essentially hand-crafted weight sets between layers. Thus, weights can be optimized to make best use of the coarsely-quantized values available to them and their associated neural states on ANNA. Furthermore, the application makes no requirement for on-line adaptation (which would be difficult, with such coarse quantization).

This is not intended to be at all derogatory to ANNA – quite the reverse. The device has been engineered to a (precision) budget, with an application in mind. The benefits accrued by such close matching between problem and solution are obvious.

2.7 MOS transconductance multiplier

The MOS transconductance multiplier was the basic building block for transversal filters [32] before it became possible to implement digital filters on Digital Signal Processing (DSP) chips with their on-chip digital multipliers. In the neural context, the use of a transconductance multiplier means that *both* the weights T_{ij} and the neural state V_j are represented by analogue voltages. Analogue multipliers are classified as one-, two- or four-quadrant multipliers. With most neural networks, the neural state V_j can be restricted to a range between 0 and 1, and two-quadrant multiplication is therefore sufficient. The synaptic weights are usually stored *dynamically* using similar principles to dynamic RAMs. Permanent storage of analogue weights is much more difficult to achieve in VLSI, and specialized techniques are required here. This is such an important issue that the whole of Chapter 3 is devoted to it.

2.8 MOSFET analogue multiplier

The expression for the drain-source current I_{DS} for a MOSFET in the *linear* or *triode* region has already been given earlier in this chapter:

$$I_{DS} = \beta \left[(V_{GS} - V_T)V_{DS} - \frac{V_{DS}^2}{2} \right] \quad \text{where} \quad \beta = \frac{\mu C_{ox} W}{L}$$

This equation can be re-written as follows:

$$I_{DS} = \beta \times V_{GS} \times V_{DS} - \beta \times V_T \times V_{DS} - \beta/2 \times V_{DS}^2 \quad (2.6)$$

The first term in the above equation is the product of two analogue voltages and a constant. The two other terms are, in effect, error terms which can be taken out via a second MOSFET, as shown in Figure 2.7. For this circuit, I_1 and I_0 are given by:

$$I_1 = \beta_1 \left[(V_{GS1} - V_T)V_{DS1} - \frac{V_{DS1}^2}{2} \right] \quad (2.7)$$

$$I_0 = \beta_0 \left[(V_{GS0} - V_T)V_{DS0} - \frac{V_{DS0}^2}{2} \right] \quad (2.8)$$

If $W_1/L_1 = W_0/L_0$ (i.e. $\beta = \beta_1 = \beta_0$) and $V_{DS1} = V_{DS0}$ then:

$$I_P = I_1 - I_0 = \beta(V_{GS1} - V_{GS0})V_{DS} = \beta(V_1 - V_0)V_{DS} \quad (2.9)$$

Hence $I_P = \beta V_X V_Y$ where $V_X = V_{GS1} - V_{GS0} = T_{ij}$ and $V_Y = V_{DS} = V_j$, for example.

The resultant product I_P is a current; the currents from different T_{ij} multiplications can be summed together at the input of an op-amp which then gives an output voltage proportional to $\Sigma T_{ij} V_j$. We will consider this circuit in more detail in the context of pulse-stream technology in Chapter 5.

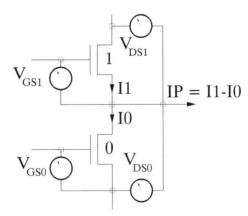

Figure 2.7. *Transconductance multiplier – yielding the product of two voltages as a current.*

2.9 Imprecise low-area 'multiplier'

Let us return to equation (2.6).

$$I_{DS} = \beta \times V_{GS} \times V_{DS} - \beta \times V_T \times V_{DS} - \beta/2 \times V_{DS}^2$$

We could choose to *ignore* the last two terms in this equation on the basis that, with massive parallelism, the exact transfer function of the processors is not critical. Akers and his team [17] use the synapse circuit shown in Figure 2.8 to implement limited interconnect structures. In this circuit, T_1 allows a value representing T_{ij} to be stored dynamically on the gate of T_2. The 'multiplication' is performed as the drain terminal of T_2 is set to a voltage representing V_j. The clock signals allow the charge packet produced by T_2's drain-source current to be passed to the analogue adder, to give a naturally thresholded V_{out}. This is not exact multiplication, but it has been claimed that the imperfections either do not matter or can be accounted for during learning [17].

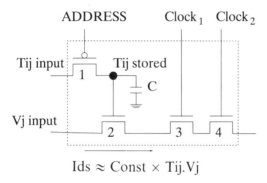

Figure 2.8. *Low-precision (non-linear) analogue multiplier based on a single MOSFET.*

2.10 Analogue, programmable – Intel Electronically-Trainable Artificial Neural Network (ETANN) chip

We will return to ETANN [33] when discussing memory technology in the next chapter. However, it is worth reviewing ETANN's methods and specifications here, in the context of this discussion of neural implementation *per se*.

Based around the well-understood Gilbert Multiplier mentioned in Section 2.5, ETANN offers 64 neurons, with up to 128 inputs per neuron. Weights are non-volatile, and have an estimated 'precision' rated at 6–8 bits, when programmed carefully. Multi-ETANN boards are available to implement larger networks, and computation speed is impressive at an estimated 2.5 billion CPS – similar to CNAPS' digital performance.

Anecdotal evidence from ETANN users indicates that the device is fickle, however – as might be expected from a chip using unusual weight technology (or at least EEPROM technology being used in an unusual context). Programming is slow, and achieving the rated weight precision non-trivial. However, ETANN is an impressive beast, cramming a great deal of $\Sigma T_{ij} V_j$ on to a chip. Its raw performance exceeds that of our own EPSILON chip, but we believe that EPSILON has some 'hidden' merits, as we will show later in Chapter 5.

2.11 Conclusions

It is not our intention in this book to promote any technique, certainly not our own, as **the best** for all neural implementation. However, it is our view that, for a number of real-time applications, massively parallel analogue (or pseudo-analogue) programmable hardware will become the optimal approach for the implementation of artificial neural networks. Furthermore, it is difficult to avoid the view that any *neural* architecture or algorithm that actually *requires* the precision and accuracy of digital hardware is not all it should be. This view cannot be substantiated by any more rational argument than the fact that the biological neural system is not renowned for its arithmetic precision. According to Hopfield [34], '...*the need for high accuracy ... is an artefact of our present understanding of learning systems, and not an intrinsic limitation ...*'. One of the driving forces behind neural computation has always been the aspiration that massive parallelism will trade off against accuracy, and that fault tolerance will ensue.

Within the analogue domain, we believe that non-volatile analogue memory will become essential to applications where programmability is important, and weights are changed relatively infrequently, by rather invasive means. Truly dynamic synaptic plasticity (i.e. on-chip learning), on the other hand, will be better served by dynamic weight storage, where charge values can be altered readily by learning circuitry, essentially 'on-the-fly'. Analogue memory forms the subject of Chapter 3, whilst the more speculative, and fascinating, topic of on-chip learning is left to the end of this book.

3

Analogue synaptic weight storage

3.1 Introduction

The requirement for suitable weight storage techniques is common to all analogue neural VLSI whether it be sub-threshold, transconductance multipliers or pulse-stream circuitry. Although digital DRAMs have now reached very high levels of density, standard cells fabricated in a standard CMOS process are not small. Digital weight storage would, therefore, occupy much of the synapse cell area, as became clear, for example, in the early pulse-stream work (Chapter 4) where 5-bit signed digital weights were incorporated. Even with this relatively low weight precision, the resultant chips could only implement tens, rather than hundreds or thousands of neurons, and this is likely to remain the case for networks using *digital* weight storage.

It is our view, although it is not a universally held one, that digital weight storage in analogue networks is not a good combination. Although the digital memory technique offers much in terms of design simplicity and process independence, it nullifies much of the gain made in using compact analogue computation circuitry. This chapter, therefore, concentrates instead on the possibility of integrating analogue weight memory, on both standard and non-standard CMOS processes. Analogue weight values may be represented by resistance, charge or voltage values, via the diverse methods described in the ensuing sections.

3.2 Dynamic weight storage

An analogue weight value can be represented by a charge value Q stored on a capacitor C, and trusted to stay there, representing a voltage $V = Q \div C$. In principle, the only restriction on the values that V may assume is that it remains within the voltage range defined by the chip's power supplies, Vss and VDD (commonly 0 V and 5 V, respectively). In common with the similar

DRAM method of digital storage, this method is subject to problems of leakage and data corruption, and requires refresh circuitry if long hold time and/or accuracy is required. Leakage of charge occurs into the substrate via the reverse-biassed source → substrate and drain → substrate diodes, and through subthreshold conduction. Data corruption will be induced through noise and internodal coupling. It is often difficult to make an *a priori* assessment of subthreshold leakage effects, as electrical design rules are generally not very forthcoming about 'leakage', preferring to assert that leakage is 'negligible'. Indeed, leakage currents are typically in the pico-amp (10^{-12} Amp) range, for a well turned-off MOSFET. The implication of this order of magnitude is that a 2.5 V weight value, stored on a typical 0.1 pF capacitor, will be degraded by 1% in 2.5 ms. This is a relatively long time, in terms of typical VLSI clocking speeds, but subthreshold leakage is not the only leakage mechanism. Currents through the reverse-biassed diodes will be of the same order of magnitude. If suitably frequent refresh is not acceptable, these currents can be reduced by orders of magnitude, and the storage times consequently enhanced, by cooling the silicon. Whether this is necessary depends critically upon the inherent process leakage, and thus on the dopant densities, impurity concentrations and defect occurrences that determine leakage levels. One implementation using dynamic weight storage claims hold times of 5 minutes (with less than 1% leakage) at room temperature, rising to several days with lowered temperatures [35].

We are experimenting with charge-balancing techniques [36] to achieve partial cancellation of leakage terms. If the stored weight can be represented by the **difference** between two stored voltages on capacitors, and both decay via similar leakage mechanisms, some level of leakage immunity is gained. Whether this is a useful added complication, enhancing the performance of the device, or merely a means of gaining a small increase in hold time, remains to be seen.

There is no *fundamental* problem with dynamic weight storage and, as will be discussed in the final chapter, it can offer advantages for on-chip learning. However, refresh circuitry is both inconvenient and clumsy. Several groups, including that at Edinburgh are considering nonvolatile, analogue weight storage – see for instance [37, 38]. The EEPROM-based MNOS work from Lincoln Labs is a good example of this approach.

3.3 MNOS (Metal Nitride Oxide Silicon) networks

This work uses essentially EEPROM (Electrically Erasable Programmable Read Only Memory) technology to store analogue weights as charge in a nitride layer between the gate and channel of a MOS transistor, thereby causing a modulation of the gate voltage [39]. The structure is shown in Figure 3.1. The normal cross-section of a MOS device is retrieved by omitting the nitride layer. In a MOS device, the potential in the 'gate' area

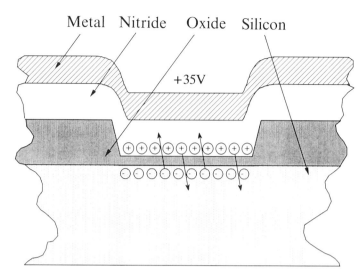

Figure 3.1. *NMOS device structure indicating schematically the tunnelling process that imbeds charge in the nitride layer.*

(under the thin oxide) is modulated by applying a voltage externally to the 'metal' (in fact, usually polysilicon – a highly conducting polycrystalline material). This applied potential can either draw carriers into the channel area under the metal gate, thus increasing its conductivity (accumulation) or drive them away (depletion), rendering the channel less conductive. In an MNOS device, with the nitride layer interposed, the metal gate is not able to influence the channel area in the same way. However, at gate voltages of $\simeq \pm 35$ V (well in excess of normal CMOS operating voltages, and at levels that would destroy any normal CMOS device via diode breakdown), large vertical electric fields appear across the oxide layer. Quantum mechanical tunnelling is induced, and electrons and holes move between the underlying silicon substrate and long lifetime trapping sites in the nitride, **through the insulating oxide layer**. This is indicated schematically in Figure 3.1. The effect of this migration of carriers into the nitride layer is to interpose a permanent 'gate' between the metal gate and the channel area. This determines the potential distribution beneath the gate, which will remain fixed until another large voltage is applied to 're-programme' it. This potential well beneath the gate of the MNOS device is then used as a reservoir of adjustable depth, filled by presynaptic elements when $V_j = 1$. In other words, the multiplication $T_{ij}V_j$ is performed by filling a well of depth T_{ij} under control of V_j. This reservoir then provides a metered packet of charge representing $T_{ij}V_j$ to be passed through subsequent CCD-like structures for postsynaptic summation. The nitride charge layer modulates the depth of the reservoir, and thus the multiplier T_{ij}. Figure 3.2 illustrates

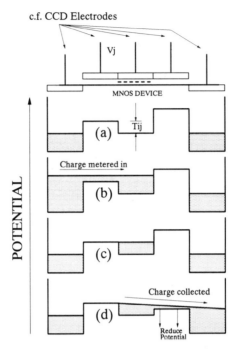

Figure 3.2. *NMOS device as an element in a CCD structure for synaptic multiplication.*

the complete MNOS/CCD structure. Charge is metered into the MNOS potential well as V_j lowers the height of the barrier between the charge source on the left, and the MNOS well, whose depth represents T_{ij} as determined by the trapped electrons/holes. Subsequent modulation of the potential barrier between the MNOS device and the synaptic output by a separate CCD clock signal allows the charge packet representing $T_{ij}V_j$ to be 'poured' into the postsynaptic summation. The network dynamics evolve at the CCD clock rate (over 10 MHz), which controls the rate at which packets are passed through the synapses. As usual, an increase in speed implies that charge packets be made smaller, with a concomitant decrease in accuracy. The raw accuracy of the MNOS synapse is determined by the accuracy and linearity of the relationship between the MNOS well depth and the applied $\simeq 35$ V programming voltage.

The MNOS technique affords *electronic* programmability, at the expense of the requirement for a non-standard process, which interposes additional process steps between those of a normal MOS process. The following paragraphs describe work involving more-or-less standard CMOS processes, and using cleverness at the subcircuit level to achieve neural function at minimal cost in transistor count.

3.4 Floating-gate technology

Examples exist which utilize another EEPROM technology – that of floating gate. Here, tunnelling is again induced in a MOS structure by the application of unusually high voltages, but through an abnormally thin area of device gate oxide. Electrons thus embed themselves in an otherwise 'floating' gate, and act in much the same way as do the electrons in the MNOS device described above. Examples can be found in the work of Vittoz [40], and in the Intel ETANN chip [33]. The principle is simple, although its execution is not. An electrode analogous to the MOSFET gate is embedded in the usual place, separated from the 'channel' underneath by thin oxide, but with no direct connections to other signals. The floating gate has, however, an exceptionally thin area of oxide separating it from an area of fully diffused substrate. Figure 3.3 illustrates the above schematically. A carefully controlled tunnelling mechanism is used, to embed carriers (electrons) in this 'floating gate', where they remain, modulating its potential, and allowing it to be used as a semi-permanent gate, clamped to a fixed potential until reprogrammed. The floating gate has the same physical effect on the level of inversion in the channel as does a normal polysilicon MOSFET gate. The (floating) gate voltage is, however, determined by the charge transferred onto the floating gate by the tunnelling induced during programming, and therefore remains non-volatile until reprogrammed. Accuracy is impressive (at least 8 bits' worth [33] when an incremental programming style is adopted) and hold times are of the order of years. Incremental programming involves setting the weight to a nominal value, and subsequently incrementing or decrementing its value via (off-chip) feedback circuitry to 'home in' on the desired actual value.

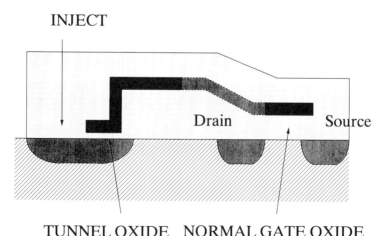

Figure 3.3. *Floating-gate EEPROM structure.*

The disadvantages of floating-gate technology are again that it is non-standard and interferes with the normal CMOS fabrication cycle, and that the programming mechanism is relatively slow. A single weight update takes several milliseconds, and the weight update cycle for a 10,000 synapse chip is of the order of many seconds. This is not a major problem during neural computation for which weights are merely programmed at the outset of a task. However, with any analogue memory technique that does not have 100% accuracy and repeatability (i.e. **all** analogue memory techniques), the chip must take part in the learning cycle. This 'chip-in-the-loop' strategy is essential to ensure that variations in memory characteristics between chips and across a chip are cancelled during learning. The penalty of the floating gate programming time is therefore a serious limitation. In effect, every individual ETANN device must be trained individually – a very time-consuming affair. This is not a problem unique to ETANN, of course. Any idiosyncracies in weight and arithmetic circuitry imply that simply down-loading a set of weights is unlikely to produce a satisfactory result. We look at techniques for generating **robust** weight sets towards the end of this book, but it will probably always be necessary that weight sets be 'fine-tuned' via the 'chip-in-the-loop' technique [33] to get the best out of analogue hardware.

3.5 Amorphous Silicon (α-Si) synapses

Amorphous silicon is commonly used as a high-resistivity medium for integrating moderate-to-high value resistors on CMOS substrates. Attempts have been made to use amorphous silicon as a **binary** memory device in the neural context, with 'high' and 'low' values representing the two binary values, programmable electronically [37]. However, the physical properties of amorphous silicon are much richer than that, and its potential much greater. At Edinburgh and Dundee Universities, over the past decade, amorphous silicon memory devices have been the subject of much study. During the course of work to develop a fast **binary** memory device, **analogue** memory properties were observed. These were regarded initially as a nuisance.

Figure 3.4 illustrates the device structure. It is a vertical sandwich, with metallic upper and lower terminals, and α-Si filling. The analogue nature of the device depends critically upon the choice of upper and lower metals, as well as the recipe for and the thickness of the Si itself. The two-terminal α-Si metal $\rightarrow p^+ \rightarrow$ metal junction devices exhibit non-volatile polarity dependent analogue memory switching after an initial application of a ~ 13 V ('**Forming**') pulse [38]. The forming process creates a vertical filament of conducting material, less than 1 µm in diameter, although the details of the conduction process are not yet fully established. It seems likely that some form of hopping between conducting areas along the filament path is responsible. Subsequent voltages in the range ± 2 to ± 5 V programme the

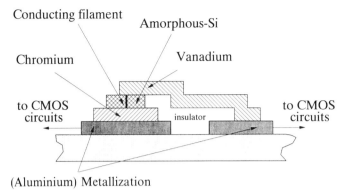

Figure 3.4. *Amorphous silicon, non-volative memory device – a vanadium-amorphous-silicon-chromium sandwich.*

device resistance *to a value determined by the magnitude and polarity of the programming voltage*. This value is held essentially indefinitely, and 'cycling' the device (repeated programming and erasing) does not appear to introduce degradation in properties. Programmed resistance values are in the range $k\Omega$ to $M\Omega$, and while non-linear, are achievable accurately and rapidly (programming pulses are nS wide). These devices are superior to existing non-volatile technologies such as MNOS and floating gate technology, in terms of speed (<1 ns), programming voltages ($<=5$ V), retention time (>4 years) and cycling ($>10^6$ cycles) [38]. Figure 3.5 shows a typical programming curve for one of these devices. The device resistance can be set to any value on the curve, by application of the appropriate programming voltage. Accuracy can be enhanced, as in the case of floating-gate technology, by adopting an incremental programming approach, whereby

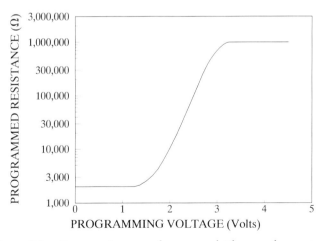

Figure 3.5. *Programming curve for a non-volatile amorphous memory.*

the device 'creeps up' on its intended resistance value over the space of several programming steps, rather than being set in a single step. This curve is typical of a Chromium/α-Si/Vanadium device. Different metals produce less 'analogue' results – i.e. the curve is more step-like.

There is some interesting physics behind this behaviour. We have also shown that the current-voltage characteristics at low temperatures exhibit step-like features, suggesting quantized ballistic transport. The incrementally variable characteristics of the device allow programming to be performed extremely accurately by applying feedback. We have investigated the accuracy and repeatability of the programming process in α-Si memories, and also attempted to lower the forming voltage, as far as 11–12 V. This is still sufficiently high to be a major problem for any standard CMOS process. We have therefore designed a 5 μm NMOS device, capable of withstanding up to 25 V, to integrate an array of α-Si devices on a MOS substrate. This is being fabricated in the Edinburgh Microfabrication Facility. We have also established that the α-Si processing steps do not affect the underlying MOS devices adversely, and we have verified that the α-Si devices work, *at least as digital memories*, on a MOS substrate.

Future work on materials and device fabrication covers both fundamental and device-oriented aspects, and will lead to full CMOS/α-Si VLSI implementations. To impart some feel for what the likely problems and options are in using such an unusual device, we present below some of the areas we are currently exploring to optimize the performance and characteristics of the α-Si devices.

3.5.1 Forming at higher temperatures

The forming stage of the memory devices currently requires higher voltages than those of normal CMOS operation. It is clearly highly desirable that we reduce the forming voltage, or even make forming unnecessary. We have already taken multi-layers of metal/α-Si and heat-treated them to 500°C prior to deposition of top metal. These devices form at much lower voltage levels than un-annealed devices. We are therefore investigating the effect of annealing, both prior to and after deposition of the top metal, on the resistance, forming voltage and switching characteristics, and the effect of forming at elevated temperatures, for a range of top and bottom electrode combinations. In fact, early samples fabricated without any such interference on 1.5 μm CMOS substrates have a significantly **lower** forming voltage ($\simeq 8$ V). The reason for this fortuitous change is not at all clear!

3.5.2 Deposition of metal during α-Si growth

It is clear from previous work that metal is incorporated into the α-Si during forming. We therefore intend to modify deposition to allow co-deposition of a range of metals such as W, Cr, V, etc. from fluoride and carbonyl gases. We will study two basic structures: homogeneously during deposition of the α-Si layer for a range of metal concentrations; and in a multi-layer metal/α-Si/metal/α-Si, etc. structure.

3.5.3 Investigation of the forming process

We must study the effect of the forming process on memory performance. With a large number of devices, we will obtain further information on the effect of fabrication and material parameters, of device thickness and pore diameter (i.e. the geometry of the device), and of parameters relating to the programming process itself.

3.5.4 Programming technology

We will have to design new switching hardware and software with emphasis on the programming accuracy and a 'feedback' loop for checking the resistance states – particularly since the α-Si devices are now to be programmed **through** MOSFETS. We must minimize the effects of external (programming circuitry) factors in determining the reproducibility of programming.

This seems a suitable note on which to end this chapter. We have followed the developers of MNOS and floating gate technologies, to develop an exciting and completely novel memory device for neural networks. However, we have opened up something of a physical and technological Pandora's box of problems, partial solutions and interesting sidelines. This seems to be the fate of all attempts to better what is already available. Dynamic storage is understandable, implementable but requires refresh. MNOS was very promising, and seems to have foundered on the rock of lack of funding – a not surprising occurrence for an expensive and speculative programme, but tragic nonetheless. Floating gate is a reasonably mature technology, owing to its EEPROM ancestry. Programming times are, however, unacceptably long. It is this background that makes the α-Si option attractive, despite its immaturity. Hopefully, we will shortly be reporting well-characterised α-Si neural networks, with 100% yield, and >10,000 synapses/chip.

Time will tell.

4

The pulse stream technique

4.1 Introduction

Analogue VLSI gives us the potential to build massively parallel arrays of interconnected neurons as tens of thousands of synapses can be integrated on a single chip. However, the shortcomings of analogue techniques are well-known and have already been discussed. Some key features of neural processing make the intrinsic limitations of analogue circuitry less of a drawback. These considerations have led us to adopt a **pulse stream** encoding technique which performs analogue multiplication under digital control. This approach lends itself naturally to continuous, asynchronous computation, as in the conceptually simple, but theoretically rich feedback networks introduced by Hopfield [2]. It is also possible to use the *same building blocks* in single or multiple layer feedforward networks, and other neural architectures.

Before beginning a detailed study of the pulse stream technique, and its many variants, let us look briefly at its origins. The pulse concept is not new – the biological nervous system has been operating on just such a principle for rather a long time – for instance [41]. Furthermore, pulse-coding in electrical circuitry is a well-established technique, as we shall see in Section 4.2. We stumbled on the pulse technique as an expedient, which enabled us to develop essentially analogue VLSI devices on a digital CMOS process. Its many merits have since emerged and established themselves as our experience in the pulse-stream area has grown. Initially, we observed that:

- Digital processes do not incorporate 'good' analogue components such as resistors and capacitors. Furthermore, transistor characteristics are not closely controlled, beyond that which is necessary to maintain correct **digital behaviour**.
- This does not facilitate conventional analogue circuit design.
- Oscillators are easy to design.

Naively at first, we drew together the above observations and began developing pulsed neurons and crude pulse-multiplication synapses. Since these tentative beginnings, we have developed a wide variety of pulse-stream neurons, synapses and ancillary circuitry. We have at no time been motivated by a desire to model the nervous system directly, in an attempt to understand its functionality better, although other workers have since taken this approach [42, 23]. We are, however, at all times influenced by the working biological exemplar, and have occasionally been guided directly by its methods – for instance, observations relating to bats [43] led to the asynchronous data multiplexing scheme described later in this chapter [44].

Pulse techniques have now been adopted by several groups [29, 42, 45–58], with a diversity of detailed approach that is both surprising and gratifying. This chapter aims to present the options with respect to pulse stream arithmetic and coding **at the microscopic (neuron/synapse) level**, including those considered to be non-optimal. In this way, it is hoped that the interested reader will be equipped to concoct his or her own pulse stream variant, optimized for a particular application and set of circumstances.

4.2 Pulse encoding of information

Like any other processing system, a neural network consists of circuit elements dealing with 'computation' and 'communication'. The choice between analogue and digital implementation of these individual elements can be made independently for the different subsystems, with the corporate goal of optimization in terms of silicon area, speed, accuracy and power consumption, etc. **for the whole network**. Furthermore, since it is now commonplace to mix analogue and digital circuits on the same chip, hybrid integrated VLSI neural networks are a practical proposition. Chapter 2 presented examples that argue persuasively for an essentially analogue computational core. In this chapter, we will also show how a pseudo-digital communication system can provide reliable exchange of information at acceptable speed – the area where analogue systems are normally weak. The proposed technique does not quantize information explicitly, and therefore preserves the high resolution of analogue processing. However, it communicates by exchange of binary pulses with fixed amplitude, thus exploiting many of the benefits of digital circuitry.

As indicated earlier, pulse encoding of electronic information is not a new idea. Communications systems have used Pulse Amplitude Modulation, Pulse Width Modulation and Pulse Code Modulation for data transmission for some time. In fact, a variant of Pulse Code Modulation signalling was introduced in 1987, as the conversion principle behind *Bitstream* D/A and A/D converters by Phillips, reducing binary samples (e.g. 16 bit) to a much faster single bit coded signal. This technique now underlies the converters in most superior Compact Disc players. It is not appropriate here to present a lengthy review of PCM techniques, as the only relevance to pulse-stream neural VLSI is that some of the merits of PCM – robustness, ease of regeneration, the ability to time-multiplex – carry through into the neural domain, as we will show. Horowitz and Hill [59] present a condensed review of pulsed techniques in communications, while Stark and Tuteur [60, pp. 134–183] provide a more comprehensive treatment. The factors that are relevant here are drawn out in the following discussion.

Pulse stream encoding was first used and reported in the context of neural integration in 1987 [61, 62]. The underlying rationale is simple:

- Analogue computation is attractive in neural VLSI, for reasons of compactness, potential speed, asynchronousness and lack of quantization effects.
- Analogue signals are far from robust against noise and interference, are susceptible to process variations between devices, and are not robust against the rigours of inter-chip communication.
- Digital silicon processing is more readily available than analogue.
- Digital signals are robust, easily transmitted and regenerated, and fast.
- Digital multiplication is area- and power-hungry.

These considerations all encourage a hybrid approach, blending the merits of both digital and analogue technology. Such a hybrid scheme must be constructed carefully, however, as examples abound where the disadvantages of two techniques have been combined to produce a seriously non-optimal system. The pulse-stream technique uses *digital* signals to carry information and control *analogue* circuitry, while storing further *analogue* information on the time axis, as will be described below. A number of possible techniques exist, for coding a neural state $0 < V_i < 1$ on to a pulsed waveform V_i with frequency v_i, amplitude A_i and pulse width δ_i. A representative selection of these is illustrated in Figure 4.1, where a time-

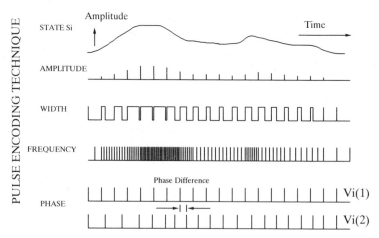

Figure 4.1. *Methods for encoding a time-varying analogue neural state onto a pulsed signal.*

varying analogue state signal V_i has been encoded in each of the following ways: **Pulse Amplitude Modulation, Pulse Width Modulation, Pulse Frequency Modulation** (average repetition rate of pulses), and **Pulse Phase Modulation** (delay between two pulses on different lines)

In addition, further variants exist – **Pulse Code Modulation** (weighted bits) and **Pulse Density Modulation** (delay between a pulse pair on the same line). These are not particularly useful in this context, and furthermore we have no direct experience in their use.

Pulse Width, Pulse Frequency, Pulse Phase and Pulse Density Modulation all encode information **in the time domain**, and can be viewed as variants of pulse rate. In other words, Pulse Density and Pulse Phase Modulation are essentially equivalent to Pulse Frequency Modulation with no averaging over pulse intervals. Such non-averaged pulse modulation techniques are used in various natural systems, particularly where spatial or temporal information is derived from the time of arrival of individual pulses or pulse pairs [43].

4.2.1 Pulse Amplitude Modulation

Here the pulsed waveform amplitude A_i ($V_i = A_i \times$ *constant frequency pulsed signal*) is modulated in time, reflecting the variation in V_i. This technique, useful when signals are to be multiplexed on to a single line, and can be interleaved is not particularly satisfactory as a long-range signalling scheme in neural nets. It incurs the disadvantages in robustness and susceptibility to processing variations inherent in analogue VLSI, as information is transmitted as analogue voltage levels.

4.2.2 Pulse Width Modulation

This technique is similarly straightforward, representing the instantaneous value of the state V_i as the **width** of individual digital pulses in V_i. The advantages of a hybrid scheme now become apparent, as no analogue voltage is present in the signal, and information is coded as described along the time axis. This signal is therefore robust, and furthermore can be decoded to yield an analogue value by integration. The constant frequency of signalling means that either the leading or trailing edges of neural state signals all occur simultaneously unless steps are taken to prevent this. In massively parallel neural VLSI, this synchronism represents a drawback, as current will be drawn on the supply lines by all the neurons (and synapses) simultaneously, with no averaging effect. Power supply lines must therefore be oversized to cope with the high instantaneous currents involved. Alternatively, individual neural signals can be staggered in time, to keep edges out of phase. This latter approach does, however, incur a disadvantage in complexity, and in the imposition of an artificial synchronism.

4.2.3 Pulse Frequency Modulation

Here, the instantaneous value of the state V_i is represented as the instantaneous frequency of digital pulses in V_i whose widths are equal. Again, the hybrid scheme shows its value, for the same reasons as described above for Pulse Width Modulation. The variable signalling frequency skews both the leading and trailing edges of neural state signals, and avoids the massive transient demand on supply lines. The power requirement is therefore averaged in time. On the negative side, however, using frequency as a coding medium implies that several pulses must occur before the associated frequency can be inferred. This apparent drawback is, however, not fundamental, as the neural state information is captured (albeit crudely) in the time domain as the inverse of the inter-pulse time. We will return to this observation.

4.2.4 Phase or Delay Modulation

In this final example, two signals are used to represent each neural state, and the instantaneous value of V_i is represented as the *phase difference* between the two waveforms – in other words by modulating the delay between the occurrence of two pulses on one or two wires.[1] This technique enjoys many of the advantages of the Pulse Width Modulation and Pulse Code Modulation variants described above, but it does imply the use of two wires for signalling, unless one of the signals is a globally-distributed reference pulse waveform. If this choice is made, however, signal skew becomes a problem, distorting the phase information across a large device, and between devices. In a massively parallel analogue VLSI device, signal skew is a fact of life.

In summary, pulsed techniques can code information across several pulses or within a single pulse. The former enjoys an advantage in terms of accuracy, while the second sacrifices accuracy for increased bandwidth.

4.2.5 Noise, robustness, accuracy and speed

The four techniques described above share some attributes, but differ in several respects from each other and in most respects from other techniques. All the techniques except Pulse Amplitude Modulation are only really susceptible to FM (or edge-jitter) noise, which will be less significant in a conventionally noisy environment. For instance, a small degradation in the voltage representing a logical 1 will only have a small effect on a system

---

[1] In biological neural systems, decisions are often made on a timescale comparable with the inter-pulse time, thus implying that a time-domain, rather than frequency-domain process is at work [43]. While this does not of itself provide justification for any technique, it is an interesting parallel.

using time as the information coding axis for analogue state, and such signals can survive transfer between devices in a multi-chip network much better than analogue (or Pulse Amplitude Modulation) signals. Furthermore, signals which are purely digital (albeit asynchronous) are easily regenerated or 'firmed up' by a digital buffer, which restores solid logic levels without altering pulse widths or frequencies. In contrast, Pulse Amplitude Modulation is neither better nor worse than straightforward Amplitude Modulation with respect to robustness.

Accuracy is, as in all analogue systems, limited by noise, rather than by the choice of word length, as it is in a digital system. We have been designing systems with 1% accuracy in synapse weights, for instance, but there is no *a priori* reason why higher or lower accuracy cannot be achieved, at the expense of more or less area. It is possible to envisage systems where a feedback, on-chip learning scheme may in part compensate for circuit inaccuracies (e.g. spread of transistor threshold). On the other hand, networks with off-line learning must rely on the precision of weights programmed into the silicon memories, or perform an elaborate pre-processing of the weights before applying them to the silicon. As we shall see in the final chapter, the issue of analogue noise in the arithmetic of an analogue VLSI device is not a simple one. Simply equating 1% noise to 6–7 bit digital (in)accuracy will not do. Digital inaccuracy imposes a quantization error on weights and states that implies that certain values are explicitly forbidden. Analogue inaccuracy, on the other hand, allows any weight/state value within the range of allowable values, but imposes a scatter of 'actual' values around that 'desired' value. The implications of this apparently subtle distinction are profound [26, 63] and as yet little appreciated.

Speed of operation is perhaps the most difficult and contentious 'figure of merit' to quantify in all essentially analogue, asynchronous systems. We can clarify the issues involved by looking at a Pulse Frequency Modulation system. The most crucial parameter is the minimum pulse width. This defines the minimum time in which anything at all can happen, and hence the maximum pulse frequency. The minimum pulse frequency is then **chosen** to give the desired dynamic range in the multiplication process. This then defines the overall speed at which 'calculations' are made. The final factor is the number of pulses over which averaging is performed, to arrive at an individual product $T_{ij}V_j$, and thus at the accumulated activity $\Sigma T_{ij}V_j$. Clearly, the technique used for multiplication, the dynamic range of the weights, and the level of activity in the network can all affect a speed calculation. We present below the results only of an attempt to estimate the speed of computation in one of the networks developed in Edinburgh, where we have been working to date on systems which require 100 pulses to drive a silicon neuron fully 'on' or 'off'. We do not present this as a serious figure of merit for our networks, but only as a guide to the sort of speed that can be achieved, basing the calculation around a modest process speed. As an example – consider a pulse-frequency system, with a **maximum** pulse frequency conservatively set at 0.5 MHz, and assume that on average, the neurons are switching from an activity level of 50–100% of the full-scale value. For a two-layer multilayer perceptron with an average level of activity (a mean neural activity of 0.25 MHz), and a typical distribution of synaptic weights (mean value, 50% of the full-scale synaptic weight value), the settling time is of the order of 1 ms, **regardless of the number of synapses on the chip.** The average computational speed may therefore be estimated as $N_s/1 \times 10^{-3}$ operations, where N_s is the number of synapses. For a typical 3600 synapse network, this equates to over 3×10^6 operations (multiply/adds) per second – a respectable number for a self-contained single chip. We treat this sort of calculation with the greatest skepticism, however, masking as it does the data dependence of the result, and the asynchronousness it implies. This is purely an averaged result, and individual calculations may take longer or shorter times.

This is but one simple example, however, and if a fast pulse-stream system is required, a move to a pulse-width technique, where the state information is encapsulated in the width of each neural pulse, will increase the speed by orders of magnitude – this approach is discussed in Chapter 5. The price paid, however, will be a loss of accuracy, as the inaccuracies in pulse edge times will no longer be integrated over several pulses, as in a Pulse Frequency Modulated system. The resultant state value will therefore have a faster response time, but will be noisier – there is always a trade-off.

What the speed calculation above does imply, however, is that pulse stream VLSI neural systems could perform many forms of computation (such as speech or image recognition) *in real time*. In truth, the speed bottleneck is likely to be shifted from pure computational speed to its normal location in analogue systems – inter-chip communication. To this end, we are developing smart intercommunication schemes that address this new limitation directly (see the end of this chapter).

4.3 Pulse stream arithmetic – addition and multiplication

There are two functions essential to the evaluation of $\Sigma T_{ij} V_j$ in a neural network – multiplication and addition. In digital systems, these are well-defined functions, although they may be implemented in detail in several ways. In analogue and pulse stream systems, there is more than one generic approach to each operation. These are discussed below, in the context of the hybrid analogue/digital pulse stream systems that are the subject of this chapter.

4.3.1 Addition of pulse stream signals

Figure 4.2 illustrates the two generic approaches to addition of pulsed (and weighted) neural states, using a frequency modulated state signal V_i as an example.

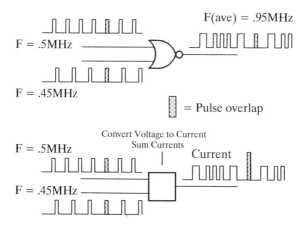

Figure 4.2. *Two generic approaches to the addition of pulsed (and weighted) neural states, using a frequency modulated state signal as an example.*

Voltage Pulse Addition (Add = Logical OR)
This technique is based on the assertion that, if the frequency of a series of fixed-width pulses (or rather their density in time – more correctly the pulsed signal's **duty cycle**) is the representational medium, performing a logical OR between two uncorrelated pulse streams is equivalent to addition of the signals they represent. In other words, if the **probability that a pulsed signal is a logical 1** is taken to represent the 'analogue' value of that signal, then ORing it with another such signal will add the probabilities, and thus, effectively, the analogue values. The underlying assumption is that pulses are by no means mutually exclusive, i.e. they can and do suffer from the overlap problem. They are, however, **statistically independent** and their overlap is thus a stochastic occurrence, and is amenable to standard statistical analysis. This OR-based add function is thus distorted by pulse overlap, which can be estimated very simply for a neural OR-based accumulator with N inputs.

ORing pulsed signals together to form the add function assumes that *the probability p_{out} that the output of the OR gate is logically high is given by the sum of the probabilities that the individual pulsed inputs are each high,*

$$p_{out}(ideal) = \sum_{0}^{N-1} p_n \qquad (4.1)$$

It is interesting to note that this immediately produces a potential nonsense, in that p_{out} can exceed 1. This observation alone should act as a warning that the OR-based technique is relying on unsound assumptions.

The *actual* probability that the OR output is logically high is the probability that one or more pulses are present at the OR gate's inputs at any instant in time, i.e. $p_{out}(actual) = 1 - p(\text{no pulses on inputs})$. This is given by:

$$p_{out}(actual) = 1 - p(\text{no pulses}) = 1 - (1 - p_0) \times (1 - p_1)...(1 - p_{N-1}) \quad (4.2)$$

The ratio between the effective 'adder' output and the 'correct' adder output is therefore:

$$\frac{p_{out}(actual)}{p_{out}(ideal)} = \frac{1 - (1 - p_0) \times (1 - p_1)...(1 - p_{N-1})}{\Sigma p_n} \quad (4.3)$$

This is plotted as a percentage in Figure 4.3, for 10, 100 and 1000 neurons, with *maximum* pulse stream mark-space ratios (which control the range of the p_n) in the range $10^{-5} - 0.5$. Clearly a 100 neuron network implemented using voltage ORing as the addition function will be less than 90% accurate unless the mark-space ratio is $\leqslant 0.005$. This means that, even for fast 20 ns pulses, the maximum allowable input pulse frequency in a 100 neuron network is around 250 kHz, thus limiting the allowable speed of calculation. The technique of ORing signals together, therefore, while it can be shown to work in small networks [61] does not scale at all well, and is not usable. Clearly, in our own early development of the pulse-stream method [61, 62, 64] we only succeeded in making this technique work because the level of integration in the test device ($\simeq 10$ neurons) and maximum mark-space ratio ($\simeq 100:1$) placed the level of accuracy at 97%.

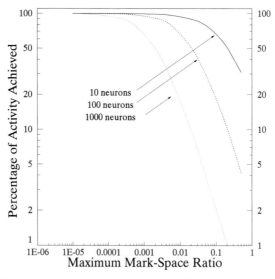

Figure 4.3. *Loss of activity information due to pulse overlap in an OR-based pulse stream adder as a function of maximum pulse mark-space ratio for 10, 100 and 1000 neurons.*

Current Pulse Addition

It is clearly possible to add pulsed voltages in the more conventional analogue way, but current pulse addition requires simpler circuitry. The same considerations relating to the rate of occurrence of pulse overlap apply here. It is their *consequences* that are different. Whereas in a voltage OR-based system, coinciding pulses register as a single pulse, in a current-based network, current pulses add, as indicated in Figure 4.2, producing a current pulse twice the size of the individual pulses. Provided the pulses are subsequently integrated in a way that preserves the integrity of this current, this does not incur any loss of activity information.

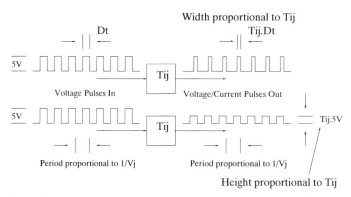

Figure 4.4. *Two generic approaches to pulse multiplication – pulse width modulation and pulse height modulation.*

Current may be accumulated either as charge on a capacitor ($V = Q/C = \int I dt/C$), or via an active integrator, based on an operational amplifier circuit, if the voltage shift in the node to which current is summed cannot be tolerated. In either case, it is clear that current summation is infinitely preferable to the voltage ORing technique alluded to above, and should be adopted for all serious pulse stream implementations. Output from current summation is no longer a pure Pulse Rate Modulated signal; it usually is averaged, and feeds a Voltage Controlled Oscillator thereby regenerating a Pulse Rate Modulated pulse stream.

4.3.2 Multiplication of pulse stream signals

This is altogether a more serious issue, about which no such straightforward and clear-cut conclusions can be drawn. As far as our experience goes, there are two generic approaches – pulse *width* modulation, and pulse *height* modulation, illustrated in Figure 4.4. We will therefore discuss the surrounding issues within this framework. The earliest pulse stream work used a simple technique that worked in the small networks in which it was tried.

We do not, however, believe that these early attempts [64] represent good use of the power of pulse stream signalling.

Synaptic gating was achieved by dividing time artificially into periods representing $\frac{1}{2}, \frac{1}{4} \ldots$ of the time, by means of 'chopping clocks', synchronous to one another, but asynchronous to neural activities. In other words, clocks are introduced with mark-space ratios of 1:1, 1:2, 1:4, etc. to define time intervals during which input pulses may be passed or blocked. These chopping clocks therefore represent binary weighted bursts of pulses. They are then 'enabled' by the appropriate bits of the synapse weights stored in digital RAM local to the synapse, to gate the appropriate *proportion* (i.e. $\frac{1}{2}, \frac{1}{4}$, $\frac{3}{4}, \ldots$) of pulses V_j to either an excitatory or inhibitory accumulator column. Multiplication takes place when the presynaptic pulse stream V_j is logically ANDed with each of the chopping clocks enabled by the bits of T_{ij}, and the resultant pulse bursts (which will not overlap one another for a single synapse) are ORed together. The result is an irregular succession of aggregated pulses at the foot of each column.

The introduction of pseudo-clock signals to subdivide time in segments is inelegant. Furthermore, ORing voltage pulses together to implement the accumulation function incurs a penalty in lost pulses due to overlap. We have therefore abandoned this technique, in favour of techniques using pulse width modulation and transconductance multiplier-based circuits. The small network introduced via this method [64], while not of intrinsic value, nevertheless served to prove that the pulse stream technique was viable, and could be used to implement networks that behaved similarly to their simulated counterparts.

Pulse Width Multiplication

When pulse frequency v_j is used to encode pre-synaptic neural state $V_j \alpha v_j$, individual pulses can be stretched or compressed in time to represent multiplication. For example, if the presynaptic pulse width is Dt_j, the postsynaptic pulse width, after passing through a synapse of weight $T_{ij} \leqslant 1$ is $T_{ij} \times Dt_j$. Pulses generated in this way represent the multiplication as postsynaptic pulses of a *width in time proportional to the synaptic weight*, and at a *frequency controlled by the presynaptic neural state*. They can therefore be accumulated in time by either of the methods described above (although we recommend current summation, for the reasons given).

Pulse Amplitude Multiplication

This is the most obvious technique, whereby pre-synaptic pulse widths and frequencies Dt_j and v_j are left unchanged, and modulation of pulse magnitude is used to perform multiplication. The resultant post-synaptic signal consists of pulses of a constant width, at a *frequency controlled by the presynaptic neural state*, and of an *amplitude proportional to the synaptic weight*. Such pulses can also be accumulated in time by either of the methods described above.

4.3.3 *Interfacing to addition*

Clearly, in either of the multiplication methods described, time- or amplitude-modulation, the postsynaptic signal must be presented to the addition circuitry in the correct form (voltage or current pulses). The manner in which this is done, along with the way in which excitatory and inhibitory synapse weights are accounted for, depends on the detailed implementation, and in particular the way in which weights $\{T_{ij}\}$ are stored. Discussion is therefore deferred until Chapter 5, where it is dealt with in the context of different exemplar synapse cells.

4.4 Pulse stream communication

We now return to the issue of communication, which is of equal importance to that of computation. When a neural network is implemented as a software simulation or using digital accelerators with a Von Neumann-like architecture, internal states are stored in memory and moved to the 'neural processors' as required. Within such architectures, moving information around in large networks does not present a fundamental problem, provided a well thought-out digital communications strategy is adopted. With analogue silicon implementations, which cast in the hardware the massive parallelism of neural systems, direct global connection is not feasible for networks with more than some tens of neurons.

Real (as opposed to 'toy') applications of neural networks require larger arrays of neurons and synapses than can be integrated on a single chip. It is therefore crucial to the effective realization of large, truly concurrent VLSI neural systems that the problem of inter-chip communication be tackled and solved now. To our knowledge, this problem has not yet been addressed directly by any group working in analogue VLSI. Some of our work, as described in the next chapter, has demonstrated that over 10,000 synapses can be included on a single chip using pulse stream techniques. However, this implies that large numbers of interneural signals have to cross chip boundaries, and allocating one or more pins to each neuron is grossly impractical. Constraints of chip pinout (even with exotic packaging technology), and wiring, both on- and off- chip, render inter-neural communication the main impediment to the implementation of large artificial neural systems.

This problem must be taken into account in the selection of the information encoding and signalling mechanism. As for any communication channel, the desired features are good noise immunity coupled with limited bandwidth requirements (i.e. adequate performance without the need for a ludicrously fast silicon process). Pulse stream techniques provide considerable benefits in terms of noise immunity, while allowing the exchange of continuous, analogue information.

These benefits are offset by poor utilisation of the *available bandwidth*,[2] unless measures are taken to improve this. In the multiplexed system described below, the scheme is designed to work with PFM (Pulse Frequency Modulated) state signals. A single PFM pulse stream consists of sparsely occurring, narrow pulses. The channel must be capable of transmitting such pulses, but it is inactive most of the time. This problem can be overcome by:

1. Multiplexing several connections over the same wire, by means of time division multiplexing. This technique can increase channel utilization at the expense of more complex interface circuitry. The main benefit is a decrease in the number of wires (or pins, for each IC), which removes or at least relieves the interconnection bottleneck for large systems. Moreover, interconnection circuits which provide the desired network connection topology are fully digital. The drawback of multiplexing is some penalty in response speed or in resolution, as discussed later in this section.
2. Designing the interface circuitry to operate much faster than the neural pulse-firing circuitry, making full use of the available bandwidth. This is equivalent in principle to using pulses whose frequency is, for example, an order of magnitude higher than that of the 'real' pulsing neurons **for intercommunication only**. The details are, of course, more complex.

Some significant criteria must be taken into account for the choice of the multiplexing technique:

- **Throughput**: the amount of information which can be transmitted in a given time (on all channels).
- **Latency**: the maximum input-output delay of information for a single channel.
- **Precision**: the (inverse of the) loss of information through the interconnection network (due to noise, quantization, etc.).

Throughput is largely a function of the available channel bandwidth and thus of the silicon technology (via the minimum pulse duration p_{min}). Latency and precision depend largely upon the details of the multiplexing technique.

Other examples of multiplexing schemes for PAM analogue neural systems exist [65, 66]. The use of time division multiplexing with *amplitude* modulated signals is not discussed here, as this technique has already been discarded by us for the reasons discussed in Section 2. In essence, two techniques for pulse stream can be envisaged for efficient time-division multiplexing of pulsed signals. A **synchronous** system uses time slots defined

---

[2] *Available bandwidth* – the maximum frequency attainable with the silicon fabrication process in use.

by a global clock, while an **asynchronous** system switches among the channels using a self-timed handshake, with no overall synchrony.

4.4.1 Asynchronous intercommunication using pulse time information

In this section we introduce a technique for communicating large numbers of neural states between chips optimally [44].

Overview

As stated above, the pulse *rate* v_i codes the neural state V_i. However, the *interpulse spacing in time* also represents an instantaneous 'snapshot' of the neural state, in a slightly different form. Information representing the state of a neuron is therefore encoded in the space between pulses δt, and it is this time interval we use to communicate essentially analogue, asynchronous information across chip boundaries via purely digital self-timed signals.

The details of the inter-chip communication system are shown in Figure 4.5. The information that crosses the inter-chip boundary is a digital pulse whose *width* represents the interval between pulses on a particular neuron i, and thus that neuron's frequency v_i and state V_i. Each neuron in turn has its state multiplexed into a purely digital, combinational circuit that codes the space between neural pulses δt as a digital pulse of length δt, from the 'sending' chip. The 'receiving' chip converts the pulse width back to an analogue voltage by using the received pulse to discharge a precharged capacitor through a constant current sink. This analogue voltage is demultiplexed on to a storage node associated with the appropriate receiving neuron, which regenerates a pulse stream with frequency v_i, inter-pulse

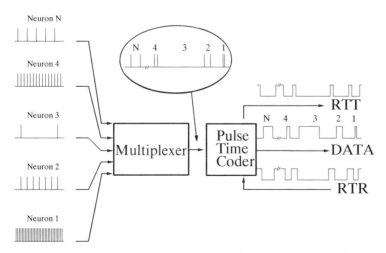

Figure 4.5. *Self-timed, asynchronous inter-chip communication scheme.*

spacing δt. A 'clone' of neuron i on the sending chip therefore appears on the receiving chip. The multiplexer's movement through the sending and receiving neuron arrays must be synchronized between chips via an asynchronous handshake pair of signals referred to here as RTT (Ready To Transmit) and RTR (Ready To Receive), as in conventional communications terminology. However, it is important to realize that **this movement through the array is completely self-timed, and the multiplexer moves on to the next neuron in the column as soon as possible.**

Details

The problem of time encoding a state $V_i \approx 0$, i.e. when no pulses, or very few pulses are present is overcome using a time-out mechanism. When a time-out is detected, the DATA line is automatically set high, to prevent indefinite wait, as is shown for neuron 3 in Figure 4.5.

The handshake lines RTT and RTR control the transmission of data between the two chips. The RTT signal is generated by the transmitting chip as a request to send data to the receiving chip. The RTR signal is taken low when the receiving chip is ready to receive data. Only once this pair of events has occurred will the transmitter send the time encoded DATA signal. If, however, the state V_i is approximately equal to 0, then the receiver driver sends RTR high, RTT is removed and the process starts again for the next neuron.

This scheme can be extended to allow one chip to communicate to several other chips by using a wired-OR arrangement on the RTR signal. Only when all chips are ready to receive DATA will data be sent, and only when all chips have received the data will RTR at the transmitting chip go high. The only condition for communication is that the transmitting chip has the same number of transmitting neurons as the receiving chip has receiving neurons, and that each chip has a common RESET line.

Co-operating state machines are used in the transmitting and receiving chips to control the interface. Each state machine is synchronous, but the transmitting and receiving state machines do not require the same clock signal. This has advantages when the chips are physically distant from each other.

Remarks

As in any multiplexing scheme, some loss of precision is implied whenever signals are compressed on to a single data line. Here, the loss of precision is determined by the value chosen for 'time-out'. This determines the lowest level of V_i that is regarded as non-zero for purposes of multiplexing, and thus affects the dynamic range of neural state. The time taken to multiplex between separate channels always represents a source of information loss in multiplexed systems. However, our self-timed scheme achieves near-optimal usage of the channel bandwidth by transmitting $\{V_j\}$ information as soon as possible, with no fixed multiplex timeslot, and is able to do so at the maximum process speed allowed. For instance, with the compression ratio of 1:10 (ten neurons multiplexed on to a single communications channel), and a 1 MHz maximum pulsing frequency with a 100-pulse 'time-out' cutoff, a typical data rate across the multiplexed channel is $\simeq 10\,\text{kHz}$. In other words, each neuron transmits its state across the channel once every 105 µs. This rate allows effective transmission of calculations performed at a timescale of the order of ms (as estimated in Section 4.2) with no significant loss of information.

4.5 Conclusions

This chapter has been an attempt to describe in detail the rationale behind pulse-stream methods. We have now begun the process of applying pulse-stream technology to problems which we believe to be suited to a hardware neural network approach. This underlies Chapter 6, and is also reported in Chapter 5, as a demonstration of the capabilities of the EPSILON pulse-stream chip.

5

Pulse stream case studies

5.1 Overall introduction to case studies

It is the intention of this chapter to provide an in-depth view of the rationale behind the design decisions in a variety of case studies, with a variety of constraints imposed by the technology, the application and the target neural architecture. The chapter is therefore almost anecdotal, in an attempt to deal with these highly practical issues. We hope that in this way intending users of pulse-stream (or indeed more general analogue) neural VLSI techniques will gain as much as possible from our experience.

Sections 5.1–5.5 describe in details the circuits that went into the Edinburgh-based SADMANN test chip, and the subsequent large EPSILON demonstrator chip. Section 5.6 describes an unusual switched-capacitor pulse stream chip, designed in Edinburgh and tested in Oxford. Section 5.7 looks at a contrasting piece of work performed largely in Oxford, where single-frequency Pulse Width Modulated (PWM) signals are used to perform 'per-pulse' rapid computation. Finally, Section 5.8 presents both the final set of circuits that went into EPSILON, and the results of some neural network experiments using that device.

5.1.1 Introduction – Edinburgh SADMANN/EPSILON work

As a group which is actively involved in many aspects of neural networks research (for example, we are currently investigating learning algorithms, novel circuits and technologies, as well as applications in control, vision and optimization), we felt it was desirable to develop a flexible research tool which would function equally well as either an accelerator for conventional computers (allowing us to speed up our simulations) or as an autonomous device for 'real world' applications. Additionally, due to the variety of target applications, the neural processor should be amenable to several different network architectures. This is not to say, of course, that we view the analogue option as a straightforward 'alternative' to a digital hardware accelerator. Analogue techniques become desirable when a combination of compactness, speed and direct sensor interfacing are desirable. In many other simply speed-critical applications, a fast DSP is what is required.

More specifically, our main applications at the time of writing include, but are not restricted to, classification using a multi-layer perceptron (MLP). Whilst the back-propagation learning algorithm (and other similar 'error reduction' schemes) used for training the MLP compensates for any synaptic multiplier variation, *provided that the chip be included in the learning loop* [33], other architectures – for example Kohonen's algorithm for unsupervised clustering [13] and Hopfield and Tank's network for optimization problems [67] – require the synapses to be matched to ensure that they function correctly. Consequently, we have invested much time and effort in developing circuits which are tolerant of both variations **across a single chip** and variations **between** chips. These circuits, and the issues associated with them, are described in greater detail in the remainder of this chapter.

5.2 The EPSILON (Edinburgh Pulse-Stream Implementation of a Learning-Oriented Network) chip

Before a more detailed discussion begins, however, it is perhaps appropriate to outline the neural network 'building block' chip which has been designed by the Edinburgh group. We will return to a detailed description and characterization of the device at the end of this chapter. The subject of the above rather contrived acronym constitutes a single layer of 3600 analogue synapses, arranged as 120 inputs by 30 outputs. Pulse coding of neural states is employed, both to simplify synapse arithmetic and to facilitate inter-chip communication of states, and is available in both synchronous pulse-width modulation, and asynchronous pulse-frequency modulation. The neurons have variable transfer characteristics (the so-called activation function), and inputs can take the form of either analogue voltages or digital pulses. Synapses are fully programmable, with weights being stored dynamically at each site and periodically refreshed from off-chip RAM. The chip is fabricated in a 1.5 µm double metal CMOS process, and measures approximately 10.1×9.5 mm. It is capable of an estimated average of 0.36 billion multiplications per second.

The chip, which has been fabricated by European Silicon Structures (ES2), is fully functional. Two EPSILON devices occupy a printed circuit board which operates as a fully analogue neural accelerator, interfaced to a host computer.

5.3 Process invariant summation and multiplication – the synapse

This section details circuitry which we have designed to allow the $\Sigma T_{ij} V_j$ operation to be performed in a process invariant manner, i.e. tolerant to the variations we expect to see across and between chips in a typical CMOS process (e.g. transistor threshold voltages, capacitances, etc.) The results reported here were obtained from tests carried out on a 10×10 synaptic array with 10 on-chip neurons, fabricated using ES2's 2 µm digital process. Wherever possible, we have tried to use small (and therefore cost-effective) test chips, in addition to SPICE simulations, to verify our circuits before committing ourselves to a full-blooded demonstrator chip. Silicon always throws up some unexpected surprises, and a measured approach is always desirable – but not always possible. The EPSILON device itself, for instance, was designed against the backdrop of a changing target process, a reduced silicon budget and a tight timescale. These are not optimal conditions for the exacting process that is chip design, and it is at the very least remarkable that EPSILON is as successful as it is.

5.3.1 The transconductance multiplier

The first synapse which we designed was based on the well-known transconductance multiplier circuit, already introduced in Chapter 2. Figure 5.1 shows the synapse with transistors M1 and M2 forming the transconductance multiplier. These transistors output a current proportional to the synaptic weight voltage V_{Tij}, which is then pulsed by the switch transistor M3, controlled by the incoming neural state V_j. By integrating the resultant output current over a period of time the required $T_{ij}V_j$ multiplication is achieved. A Voltage Controlled Oscillator (VCO) then converts this activation level into a stream of fixed width pulses.

The operation of this synapse can be explained with reference to the I_{DS} equation for a MOSFET transistor in its linear region of operation – described in Chapter 2. Equation (2.2) is repeated below.

$$I_{DS} = \beta \left[V_{GS} V_{DS} - V_T V_{DS} - \frac{V_{DS}^2}{2} \right] \tag{5.1}$$

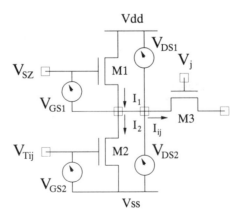

Figure 5.1. *Transconductance multiplier.*

As already pointed out, the problem with equation (5.1) is the presence of non-linear terms in addition to the $V_{GS}V_{DS}$ term. A second transistor, M1, which is identical to M2, is used to eliminate these terms. For the second and third terms of equation (5.1) to be cancelled exactly, we have already seen that the V_{DS} and V_T voltages must be the same for both transistors. While it is straightforward to arrange for V_{DS1} to equal V_{DS2}, transistors M1 and M2 have different V_{BS}. They will therefore have significantly different threshold voltages, owing to the body effect. From calculations based on standard body effect equations [19], we have found that the difference in threshold voltages is about 0.2 V. The equation for the output current of the transconductance synapse therefore becomes:

$$I_{ij} = \beta \left[((V_{GS1} - V_{T1}) - (V_{GS2} - V_{T2}))V_{DS} \right] \qquad (5.2)$$

Thus the current I_{ij} is directly proportional to the voltage difference $((V_{GS1} - V_{T1}) - (V_{GS2} - V_{T2}))$ multiplied by V_{DS}.

In this implementation the transconductance multiplier is actually being used as a voltage controlled current source, because of the pulsed nature of the pre-synaptic neural state signal. This is achieved by keeping the V_{DS} voltages constant and only varying V_{GS1} (i.e. the synaptic weight). V_{GS2} determines the value of V_{GS1} at which the output current is zero. Excitation (i.e. a positive synaptic weight T_{ij}) is achieved when $V_{GS2} > V_{GS1}$; the inequality is reversed for inhibition (i.e. a negative T_{ij}). The output of the circuit M1/M2/M3 is therefore a stream of *current* pulses whose positive or negative **amplitude** is proportional to V_{Tij} and whose **frequency** is proportional to V_j.

As Figure 5.2 shows, the neuron for the transconductance synapse is composed of an operational amplifier based 'leaky' integrator and a Voltage Controlled Oscillator (VCO). The 'leaky' integrator sums the packets of charge from a column of these transconductance synapses, converting them into the neuron's activity voltage. This voltage then controls the duty cycle of the VCO which has a sigmoidal transfer characteristic.

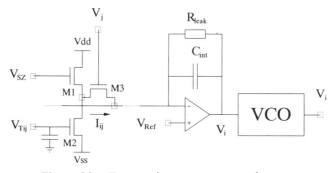

Figure 5.2. *Transconductance synapse and neuron.*

The fact that all transistors are n-types means that no well crossings are required. This, combined with the low transistor count, makes this synapse an attractive design. The resulting small area allows thousands of synapses to be implemented on a single chip.

SPICE simulations using the Level 2 transistor models for the European Silicon Structures 2 μm process, revealed three problems with this system. The first relates to the transconductance multiplier itself. The effects of process variation caused the **slope** of the synapse's voltage-to-current characteristics to vary by up to $\pm 11\%$ around its mean value. Equation (5.2) shows that variations in the surface mobility of the transistor channel μ_o and the gate oxide thickness t_{ox} have a direct effect on the magnitude of the output current via the process factor $\beta = \mu\varepsilon/t_{ox} \times W/L$. The variation in $(V_{T1} - V_{T2})$ is small at $\pm 20\,\text{mV}$ and thus in this instance can be ignored. Studying the ES2 2 μm process parameters revealed that the tolerances on μ_o cause a $\pm 5\%$ variation in this parameter with t_{ox} varying by $+4\%/-6\%$. This accounts for the overall 11% variation that surprised us so much initially.

The second problem is that the output current is sensitive to the value of the mid-point voltage between transistors M1 and M2. To maintain this point to 1% of the transistors' V_{DS} the amplifier requires to be a two stage operational amplifier. Other amplifiers, such as inverters, have neither the drive capability (high enough I_{DS}) or a high enough gain (> 1000).

The third problem relates to the 'leaky' integrator. As the number of synapses per neuron is scaled up both the maximum current and the capacitance will also increase. As the operational amplifier in the 'leaky' integrator has a finite current drive capability and is only compensated up to a specified capacitive load, every time the system is scaled the operational amplifier will need to be redesigned. This effectively renders a system based on this circuit 'uncascadable', although the problem will really only manifest itself in a neural network in pathological cases. Much of the time, the **average** synaptic current will be small – only when, for instance, all the synapses deliver a large excitatory current will the cascadability problem become acute.

This combination of factors led us to modify the design to remove this nagging worry.

5.3.2 A synapse based on distributed feedback

To solve the problems of process variation and poor cascadability a *distributed* buffer stage (transistors M4 and M5 in Figure 5.3) was added to the transconductance synapse. The operational amplifier at the foot of each synaptic column provides a feedback signal V_{outi}, which controls the current in all the buffer stages in that column. The current being sourced or sunk by the multipliers is thus correctly balanced, and the mid-point voltage is

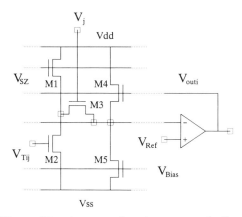

Figure 5.3. *A synapse based on op-amp feedback.*

correctly stabilised. The gate voltage of transistor M5 V_{Bias} determines the voltage level about which V_{outi} varies. The buffer stages combined with the feedback operational amplifier are functionally equivalent to a standard operational amplifier current-to-voltage converter, where the resistor in the feedback loop has been replaced by transistor M4.

To simplify the analysis of a column of N synapses, the synapse was broken down into its two components, the transconductance multiplier and the buffer stage. Assume that, when M3 is on, the voltage across it is zero. The synapse can then be analysed as a pair of back to back transconductance multipliers (equation (5.2)), where $V_{ss} = 0\,V$ and $V_{dd} = 2V_{ref}$.

Solving for V_{outi} then yields

$$V_{outi} = \frac{1}{N} \frac{\beta_{Trans}}{\beta_{Buf}} \sum_{V_j = 5V}^{N} (V_{Tij} - (V_{SZ} - V_{Ref} - (V_{T1} - V_{T2}))) \\ + V_{Bias} + V_{Ref} + (V_{T4} - V_{T5}) \qquad (5.3)$$

Therefore the output V_{outi} is dependent on a ratio of βs rather than directly on β, as was previously the case. Provided that transistors M1, M2, M4 and M5 are well matched (i.e. they are physically close to one another), the effects of variations in the surface mobility of the channel and the gate oxide thickness are cancelled to a first order. This is analogous to switched capacitor filter techniques, in which **ratios** of capacitors are used rather than absolute capacitance values.

The problem of cascadability has also been addressed by this stratagem, as the remainder of the operational amplifier, at the foot of the synaptic column, only drives transistor gates. It is the 'distributed buffer stage' which supplies the current being demanded by the transconductance synapses, doing so locally. The resulting operational amplifier is much more compact, due to the reduced current and capacitive drive demands.

However, the output voltage of the operational amplifier now represents a 'snapshot' of all the weights switched in at a particular moment in time and not the post-synaptic activity as was previously the case. This output voltage must therefore be integrated over time to form the post-synaptic activity. The design of the requisite voltage integrator is discussed in Section 5.3.4.

An on-chip feedback loop incorporating a transconductance multiplier with a reference zero weight voltage at the weight input is used to determine the value of V_{SZ} automatically. This mechanism compensates for process variation between chips.

The layout of the synapse circuit occupies an area of 130 μm by 165 μm including the weight storage capacitor and addressing transistors.

As shown in Figure 5.4, which compares pre-design simulation results with post-fabrication measurements, the predicted results from SPICE simulations correspond closely to the measured results from fabricated circuits. These results also confirm the linear relationship between the synaptic weight voltage and the output voltage of the operational amplifier. The output voltage of the operational amplifier is plotted against the weight voltage T_{ij} with transistor M3 switched on. To assess the effect of process variations, we have measured the variation in output voltage between amplifiers with the same inputs, on the same chip, and across the amplifiers' linear operating regime. The worst case variation, expressed as a fraction of the total operational amplifier output range was found to be 6%. This result is very encouraging, and compares favourably with the 20–50% variation in

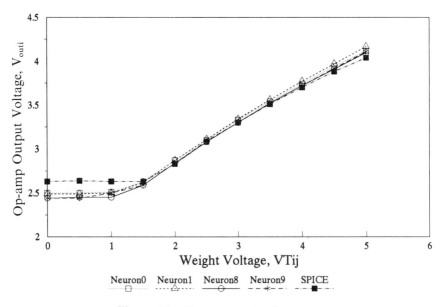

Figure 5.4. *Process variation for synapse.*

performance quoted for simple current mirrors fabricated using the MOSIS digital process – a fundamentally **digital** process analogous to our ES2 route to silicon [68, 69, 23] and the $\pm 10\%$ variations in the transconductance synapse output current mentioned previously.

5.3.3 The feedback operational amplifier

With the maximum pulse rate we have selected (0.5 MHz), the feedback amplifier needs to be fast enough to be able to respond to a 1 µs pulse width (for a mark:space ratio of 1:1). This gives a slew rate specification of 5 V/µs. The required 1% accuracy for the value of V_{Ref} translates into the need for an amplifier gain of at least 1000. It also has to be able to drive capacitive loads of up to 20 pF. A two-stage operational amplifier is required once again to meet these stringent specifications. The resulting amplifier occupies 250 µm by 165 µm.

5.3.4 A voltage integrator

Since the output voltage of this amplifier represents a 'snapshot' of all the weights switched in at a particular moment in time, the operational amplifier needs to be followed by a voltage integrator, to obtain the aggregated post-synaptic activity.

This integrator is composed of a differential stage and cascode current mirrors (Figure 5.5). The current I_{EXT} through these current mirrors is determined off-chip to minimize the effects of process variation on the integrator's output current range.

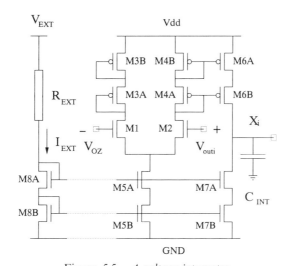

Figure 5.5. *A voltage integrator.*

The differential amplifier steers currents down the two paths, M1/M3A/M3B and M2/M4A/M4B, according to the voltage difference between the signals V_{outi} and the reference voltage V_{oz}. When these two inputs are at the same voltage, the current through transistors M3A/M3B ($I_{M3A/M3B}$) equals the current through M4A/M4B ($I_{M4A/M4B}$). $I_{M4A/M4B}$ is thus half the value of $I_{M5A/M5B}$. At this point the current being supplied to the integration capacitance should balance the current being removed. $I_{M5A/M5B}$ must therefore be mirrored on to the integration capacitor by a factor of a half, so that the net current to the integration capacitor is zero.

As both transistors M1 and M2 are in their saturation regions of operation, the actual voltage to current relationship is sigmoidal. However, by reducing the gain of the differential stage, the sigmoid function can be made to be more linear over the required input range. Since the integrator capacitor has been implemented as an NMOS transistor, any variations in the gain of the differential stage are tracked by the variations in the integration capacitance. The rate of change of voltage should therefore remain the same over all process variations. The resultant integrator occupies an area of 165 μm × 200 μm.

Figure 5.6 shows that there is a difference between the measured results and the simulated results for the voltage integrator. The problem centres on the current mirror from the differential stage load to the integration capacitor. In the fabricated voltage integrator the mirrored current is 25%

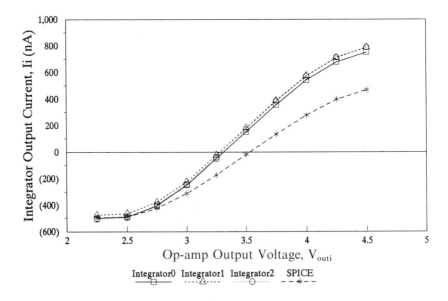

Figure 5.6. *Process variation for voltage integrator.*

too large. Frustratingly, no explanation was found to explain this mismatch problem, which occurs in results taken from a test chip. However, this problem can be circumvented by allowing the currents through transistors M5 and M7 to be varied independently. While we acknowledge that this does not address the underlying problem it does at least allow the integrator to be correctly balanced. This technique was employed in the design of EPSILON (which followed that of the SADMANN test chip), and the problem disappeared. This form of palliative measure is scientifically unsatisfying, concealing as it does the underlying cause of the mismatch. However, the desire to generate a large, working device over-rode the loftier desire to understand things fully, particularly when working against time. Figure 5.6 also shows that mirrored currents, which should be equal, actually differ by about 5–10%.

5.3.5 *The complete system*

The array fabricated as a test chip used the VLSI neurons detailed in the following sections to implement an array of 20 neurons and 200 synapses [70].

The operation of a column of synapses and the associated neuron is illustrated in the oscilloscope photograph of Figure 5.7. The top trace shows a pulse stream arriving at the synapse inputs. The synaptic weights have been set to be fully excitatory. The integrator output, shown in the second trace, increases linearly with time. As the integrator output, and therefore the neuron activity increases the neuron switches on, outputting the pulse stream shown in the third oscilloscope trace.

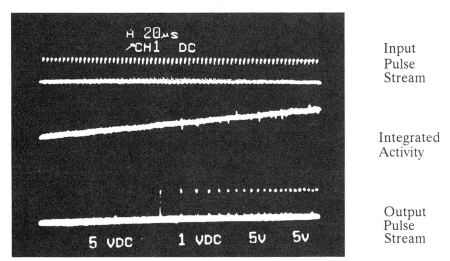

Figure 5.7. *Oscilloscope photograph of the system in operation.*

The speed of operation of this network is limited by the performance of the operational amplifier used in the synapse feedback circuit. As stated above, the limited bandwidth of the operational amplifier design employed here restricts the input pulse width to 1 µs. In other words, although the **synapse** can deal with higher bandwidths, the **neuron** is placing a speed constraint on the network.

5.4 Pulse frequency modulation neuron

As was mentioned earlier, we have employed two different types of encoding for neural states: pulse frequency modulation and pulse width modulation. In the former, the post-synaptic activity acts as an input to a Voltage Controlled Oscillator (VCO), which outputs a stream of fixed width pulses. The pulse frequency is representative of the neural state. In the pulse width modulation scheme, the post-synaptic activity acts via a comparator to produce a variable width pulse representing the neural state. The remainder of this section details two VCO circuits we have designed, whilst the subsequent section describes the pulse width modulation system.

In the VCO circuit presented here, a single capacitor is charged to determine the output pulse width, while its discharge time determines the output pulse spacing. We have developed techniques to control the charging and discharging currents in order to achieve the required circuit characteristics.

The complete circuit for a pulse stream neuron is shown in Figure 5.8. The capacitor C at the heart of the circuit is alternately charged by the current *IH*, and discharged by the current *IL*. The capacitor is followed by a comparator circuit with hysteresis. While the comparator output is LOW the capacitor is charged up by the current *IH* until it reaches the upper switching threshold of the comparator. The comparator output then changes to a HIGH state allowing current *IL* to discharge the capacitor until the lower threshold of the comparator is reached. The comparator output changes to a LOW state and the process repeats itself.

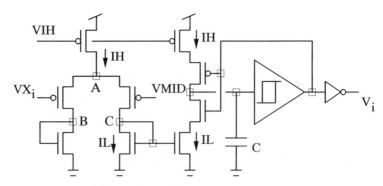

Figure 5.8. *Pulse stream neuron circuit.*

The time taken for *IH* to charge the capacitor sets the pulse width output of the neuron circuit and is controlled by the voltage VIH. The space between pulses is set by the current *IL* discharging the capacitor. The minimum pulse spacing corresponds to a duty cycle of 50%, where the pulse width equals the pulse spacing. The maximum value of *IL* is therefore equal to *IH*. The maximum pulse spacing corresponds to a zero value for *IL* and to a duty cycle of 0%.

To obtain a sigmoid transfer characteristic between input activity voltage and output duty cycle, a modified differential stage has been used to control the current *IL*. The voltage *VMID* sets the midpoint of the sigmoid characteristic corresponding to an output duty cycle of 25%, provided that one of the differential transistors is twice the size of the other. The slope of the sigmoid characteristic is set by the physical size of the transistors, and is proportional to the square root of the current flowing through this stage.

A test device comprising 10 neurons was fabricated using ES2's 2 µm CMOS technology. Each neuron circuit occupies an area of only 165 × 165 µm.

The input activity voltage to output duty cycle characteristic was measured for three neurons on each of eight devices. The results of these measurements are shown graphically in Figure 5.9. The voltage VMID was set to 2.5 V. The mean duty cycle output is plotted for each input activity value. The vertical bar represents plus and minus one standard deviation from the mean value and

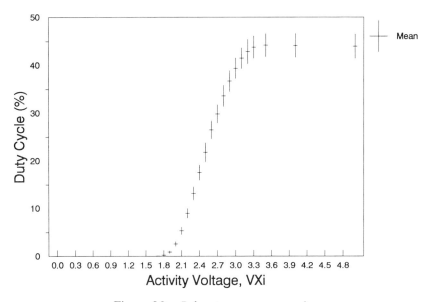

Figure 5.9. *Pulse stream neuron results.*

gives a graphical indication of the spread of results. This graph shows the predicted sigmoid activity voltage to duty cycle characteristic approximately centred on VMID. The relatively small variation in circuit performance across a sample of eight devices is also demonstrated. The spread of results is at a maximum at high values of VXi. This is due to imprecise current mirroring across the chip due to variations in threshold voltage.

To compensate for variations between chips, and to ease the design of multi-chip systems, we have developed circuitry which is used to set the pulse width output of the neuron automatically. The technique employs a phase lock loop which locks the signal from a reference neuron operating at 50% duty cycle output to match that of a reference clock generated from off-chip. In multi-chip systems, this reference signal would be distributed to all the devices. The phase lock loop controls the frequency of oscillation of the neuron by adjusting the current *IH* in Figure 5.8.

5.4.1 A pulse stream neuron with electrically adjustable gain

To have more control over network dynamics in, for example, back propagation networks, the activation function used can have programmable gain. It is possible to do this in the neuron circuit in Figure 5.8 by varying the current through the differential input stage. This has been achieved by adding extra current at node *A* in Figure 5.8. This extra current must be removed after the differential stage from nodes *B* and *C* to maintain the maximum value of *IL* used to control the pulse spacing. The current removal can be controlled using an additional differential stage as shown in Figure 5.10. This has the small advantage that the centre point of the sigmoid characteristic remains fixed, while the curve is altered around this fixed point. The gain of the sigmoid is set electrically by controlling *VIG*.

A SPICE level simulation of this circuit is shown in Figure 5.11, illustrating the change in gain of the sigmoid function for gain currents in the range 0–3*IH*.

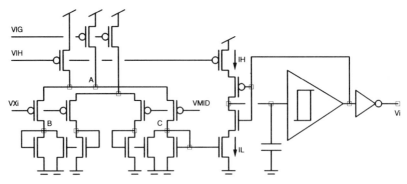

Figure 5.10. *Pulse stream neuron with variable gain – circuit.*

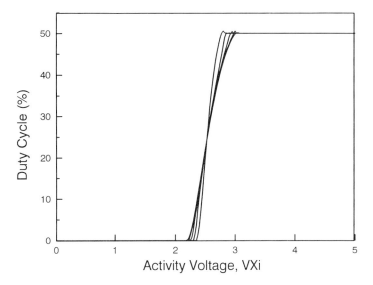

Figure 5.11. *Pulse stream neuron with variable gain – results.*

5.5 Pulse width modulation neuron

The second neuron design which we employed on our demonstrator chip used a synchronous pulse-width modulation (PWM) encoding scheme. This contrasts markedly with the *asynchronous* pulse frequency modulation (PFM) scheme we have just described.

The main motivation for the development of the PWM system was the fact that calculations using the pulse frequency method require several pulses to perform an arithmetic function. For example, a network consisting of highly excited neurons will produce a solution to a given accuracy more quickly than a network whose neurons are less excited. This may not be a major problem, but in certain speed critical applications (e.g. visual processing) a more satisfactory (i.e. less data dependent) system is required.

With this in mind, pulse width modulation was adopted. Whilst retaining the advantages of using pulses for communication/calculation, this system could *guarantee* a maximum network evaluation time. In the first instance, the main disadvantage with this technique appeared to be its synchronous nature – neurons would all be switching together causing larger power supply transients than in an asynchronous system. We have, however, circumvented this problem via a 'double-sided' pulse chopping scheme, which will be more fully explained later.

The operation of the pulse-width modulation neuron is illustrated in Figure 5.12. The neuron itself is nothing more elaborate than a two-stage comparator, with an invertor output driver stage. The inputs to the circuit

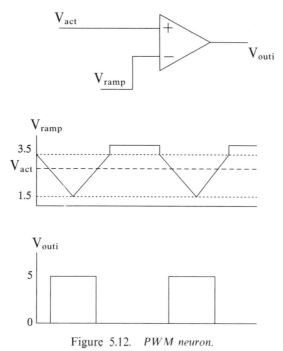

Figure 5.12. *PWM neuron.*

are the integrated post-synaptic activity voltage V_{act}, and a reference voltage V_{ramp}, which is generated off-chip and is distributed globally to all neurons in parallel. As seen from the waveforms in Figure 5.12, the output of the neuron switches whenever the reference signal crosses the activity voltage level. An output pulse, which is some function of the input activation, is thus generated. The transfer function is entirely dependent on the shape of the reference signal – when this is generated by RAM look-up table, the function can become completely arbitrary and hence user programmable. Figure 5.12 illustrates a linear transfer function – the method used to generate a sigmoidal form will be illustrated when discussing the EPSILON device, later in this chapter. The use of a 'double-sided' ramp for the reference signal was alluded to earlier – this mechanism moves both the rising and falling edges of the pulse away from their positions for maximum pulse width, thereby reducing the number of coincident edges. This pseudo-asynchronousness obviates the problem of larger switching transients on the power supplies.

The 'dead' time between successive ramps on the reference signal must be greater than the duration of the ramps. It is during this period that the integration capacitor is reset before the next set of synaptic calculations are performed and integrated. The integrated voltage is held, and the neural state signal is formed, as explained above, by the shaped ramp voltage.

This circuit is compact and simple, and as previously stated, has the advantage that network evaluation times are guaranteed. Furthermore, because the analogue element (i.e. the ramp voltage) is effectively removed from the chip, and the circuit itself merely functions in a digital manner, the system is immune to process variations.

5.6 Switched-capacitor design

In Chapter 1, we stated that switched capacitor techniques could *not* be used to introduce programmability into the summing amplifier design originally proposed by Hopfield because of the need for one VCO per synapse. In fact, the design presented in this section will show how the pulse stream concept can turn the whole problem around.

Consider the circuit shown in Figure 5.13. The feedback resistor R_f is implemented as a switched-capacitor resistor, namely as a capacitor C_L, switched by a *global* clock signal (common to all neurons) of frequency f_L Hz. As with standard switched capacitor amplifiers, capacitor C_f maintains feedback around the op-amp during the transitions of the clock signal. To make each synapse programmable, it would appear, as we have already said, that each input resistor would require a separate clock signal. With several thousand synapses in a VLSI neural network, such an approach cannot be contemplated. The pulse-stream technique, however, provides a very elegant solution to this problem. Since the neural states V_j are encoded as pulse streams, we can use them here as the *clock signals* for the switched-capacitor synapses (M_1, M_2, C_i) with the weight T_{ij} as the *input voltage* to the switched capacitor synapse. (This voltage is stored on capacitor C_T and refreshed

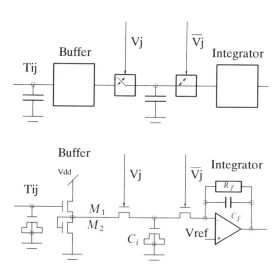

Figure 5.13. *Asynchronous switched capacitor synapse and neuron.*

periodically from an external RAM via a D/A converter.) For positive weights, the buffered synapse weight voltage is set to be less than V_{ref} and *vice versa* for negative weights. All the synapses in a column are connected to the inverting input of the op-amp to give a neural activity x_i where

$$x_i = -\frac{C_i}{C_L f_L} \Sigma T_{ij} V_j \qquad (5.4)$$

The synapse is small (65 × 65 μm), which implies that around 12,000 synapses could be integrated on a standard die. Prototype chips have been fabricated using a standard 2 μm CMOS process, on a 3 × 3 mm die, of which only 780 × 880 μm was used to implement 144 synapses and 12 neurons (i.e. >200 synapses per mm²). Digital control of the weight refresh circuitry also necessitates a small silicon overhead.

5.6.1 Weight linearity

The synapse weight voltage, stored on capacitor C_T, is buffered by an unusual two-transistor buffer circuit, consisting of an NMOS source impedance, *and an NMOS pull up device*. This circuit was used because realistic analogue VLSI neural networks will require in the order of *tens of thousands of synapses*. Area is therefore of crucial importance, and complementary device structures consume large quantities of silicon real estate. A complication of this unusual circuit form, however, is that the digital process has no facility for creating well-conditioned capacitors of moderate value. MOS devices, biased into a suitable region of operation, therefore had to be used as the storage elements. The combination of the inevitable body effect present in an NMOS device operating with a variable source potential greater than the local substrate potential and of the non-linear capacitance-voltage characteristics of the MOS storage device places a restriction on the available dynamic range of the weight voltage. The buffer transistor aspect ratios were chosen such that a reasonable approximation to a linear transfer characteristic between *applied* synapse weight and *effective* synapse weight could be obtained. The buffer transistors were designed with relatively large geometries so that the effect of length and width variations could be minimized. The storage capacitor was laid out in such a way as to minimize indirect cross-talk via capacitive coupling.

5.6.2 Weight storage time

The method of weight storage is again dynamic which brings with it the fundamental problem highlighted in Chapter 3, leakage, principally leakage through the reverse biassed source-substrate diode. SPICE2 simulations of the effects of weight leakage were at best a *very* crude approximation, since the crucial parameters required by the program are ill-conditioned and not provided by the manufacturer. The problem of weight decay in the prototype chips was analysed by monitoring the neural activity x_i in response to a weight which was programmed and then allowed to decay.

Figure 5.14 shows the effect on neural activity of 12 synapses initially programmed with $T_{ij}=0$ (which corresponds to an applied voltage of 3.5 V), each fed with a constant neural input. The average decay rate at each synapse can be approximated by dividing the neural activity rate of change by 12, and then also by the gain of the neuron, as set by the global clock frequency. This gives a decay rate at the synapse of 20–30 mV/s which can be considered to be more than acceptable even if large numbers of synapses have to be refreshed.

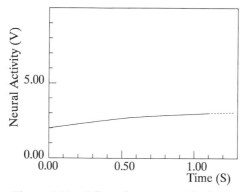

Figure 5.14. *Effect of synapse weight decay.*

5.6.3 *Accuracy of computation*

The switched-capacitor chips described in this section were tested on a pattern processing problem, as reported in Chapter 6. This enabled us to obtain an idea of the accuracy with which the $\Sigma T_{ij}V_j$ computations were being performed; with 8-bit quantized weights (i.e. using an 8-bit D/A converter to refresh the weights), the $\Sigma T_{ij}V_j$ values were measured to be within 1.2% of the values obtained for a simulation of the same pattern processing problem on a SUN workstation, also with 8-bit weights.

5.7 Per-pulse computation

This circuit was designed with multi-layer architectures in mind. Previous experience with training multi-layer perceptrons on a number of problems had shown, for example, that neural activities with these networks can be constrained to lie within 0 and 1.0. Input values to the first layer can be scaled to lie within this range without loss of performance, and the use of a sigmoid function for the non-linear activation function ensures that intermediate and output values also fall within these limits. With this design, therefore, the *width* of each pulse is used to encode the neural state between values of 0 and 1.0. The chip is driven from a single fixed-frequency square wave master clock. The first half of the clock cycle is reserved for the implementation of the synaptic multiply and add function (the *active period*). During the second half of the cycle, the result is passed through the sigmoid non-linearity and the circuitry is reset for the next computation (*reset period*). A neural state of 1.0 is simply encoded as a 1:1 square wave. The absence of any pulse (continuous logic low) denotes a value of 0, and a value of 0.5 is encoded as a pulse stream with a 1:3 mark-space ratio (i.e. a logic high for 50% of the active half of the clock cycle – see Figure 5.15). A complete multiplication and summation occurs (for every synapse-neuron layer) with every cycle of the master clock. A pipelined implementation is envisaged to overcome the problem of very large input fan-ins so that any size of problem can be accommodated with no change in throughput and only a slight 'pipeline delay'.

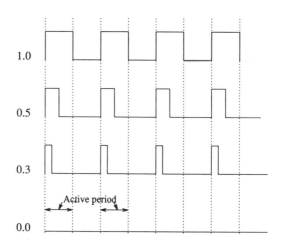

Figure 5.15. *'Clock' signals for per-pulse system.*

5.7.1 Design overview

Inputs are of the form of pulse width modulated waveforms (although direct analogue input is also possible). All waveforms are synchronized to the leading edge of the master clock. Each input is used to switch a current source/sink synapse, and the output currents of columns of synapses are summed onto a capacitor over the active period of the master clock. During the reset period, the sigmoid transfer function is applied to this voltage to generate an output voltage which is then converted back into pulse width modulated format during the next active period of the master clock. For each synapse, there is a weight capacitor refreshed asynchronously and independently of the network computation.

5.7.2 Input stage

Most multi-layer network problems require large numbers of input units, and it soon becomes impractical to allocate an input pin to each input signal. To allow the possibility of very large input fan-ins, a multiplexing system has been developed.

The chip inputs are divided into 'banks', any of which can be selected at one time. When a bank is selected, the input pins are connected straight through to the synapse array for that bank, and at the same time a capacitor (**C1** in Figure 5.16) is charged from a fixed current source for the duration of that input pulse. When a bank of inputs is unselected, an identical current is used to charge an identical reference capacitor **C2** for the duration of the active period. A pulse is generated whose width depends on the length of time for which the **C2** reference voltage is below that of **C1**. This circuitry is thus a form of 'pulse-width-sample-and-hold'.

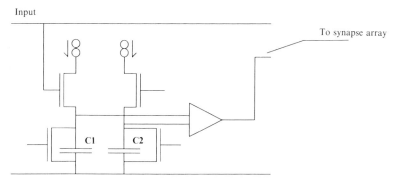

Figure 5.16. *Input stage for per-pulse system.*

5.7.3 Synapse

The synapse design is based on the three-transistor transconductance synapse already mentioned several times in this book. In this case, instead of using two NMOS transistors, a simpler approach using a PMOS and an NMOS transistor has been adopted (Figure 5.17). Here the transistors simply act as current sources, and the current difference is taken from or into the neuron during the time that the V_j input pulse stream is high. Thus multiplication is performed using the process of charge transfer, $Q_i = T_{ij} V_j$. By using CMOS transistors the synapse is robust against small changes in the mid-point voltage. The current drawn through the synapse is maintained at an almost constant level by applying the inverse of the pulse stream ($\overline{V_j}$) to

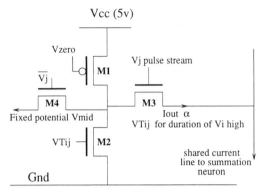

Figure 5.17. *4-transistor synapse for per-pulse system.*

a fourth transistor connected to a constant potential. This fourth transistor largely overcomes the switching problems of this circuit.

5.7.4 Summation neuron

The summation $\Sigma T_{ij} V_j$ is again generated by tying together all the synaptic currents for each neuron into an op-amp integrator. The total current is then integrated over the active period to give the total charge, Q_i (Figure 5.18). The op-amp used is a 'dynamic amplifier', i.e. an invertor biased so as to act as a high gain amplifier over the period of interest. This technique requires a very small silicon area. As the integrator must be reset every cycle (for the next computation), the dynamic biasing required for such an amplifier is provided at no extra cost.

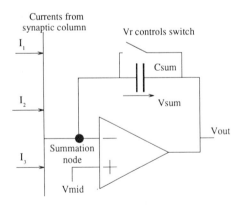

Figure 5.18. *Summation neuron for per-pulse system.*

5.7.5 Sigmoid function

A simple circuit which implements a sigmoidal activation function is the long-tail pair of Figure 5.19. For small variations of the differential input voltage a linear change of current in each branch is obtained. As the differential input $(V_{in} - V_{ref})$ is increased, the circuit saturates, eventually giving a current which tends to zero or to I_{tail}, thereby providing the required sigmoidal activation function. The current in one branch is mirrored and used to charge a capacitor for a fixed period. In this way a sigmoidal voltage conversion is produced. By externally varying the I_{tail} current, the gain of the sigmoid function in its linear range can be controlled.

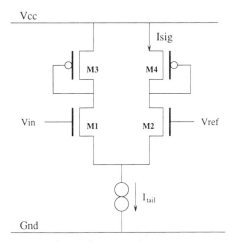

Figure 5.19. *Long-tail pair for sigmoid generation – per-pulse system.*

5.7.6 Pulse regeneration

To convert the sigmoidal output voltage back into a pulse stream, the same approach as is used for the input sample-and-hold circuit can be adopted. The voltage from the sigmoid function is compared to that of a ramp reference voltage, and an output pulse generated whose width depends on the time during which the output from the sigmoid function is greater than the reference voltage.

5.7.7 SPICE simulation

Figures 5.20–5.22 are taken from SPICE simulations extracted from the layout. In all these plots, a single synapse column and neuron were simulated for 40 μs. During the simulation, an input pulse stream equivalent to an input value of 1.0 was applied to a single synapse, whilst the synaptic

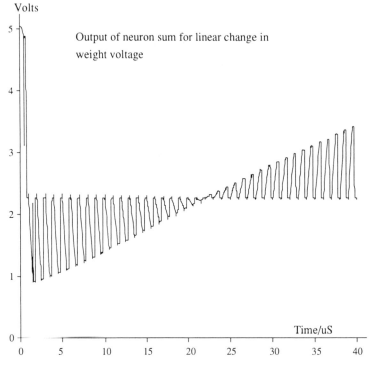

Figure 5.20. *SPICE simulation, output neuron, per-pulse system.*

weight voltage was gradually ramped up (over its linear range). Figure 5.20 shows the output of the neuron summation unit (the dynamic op-amp). It shows the neuron being reset every 1 µs by the master clock, and the near linear weight/voltage relationship of the synapse. Figure 5.21 shows the voltage at the output of the sigmoid non-linearity when the summation voltage of Figure 5.20 is applied at the input. Finally, Figure 5.22 shows the resultant output pulse stream demonstrating the pulse width coding scheme. These plots clearly demonstrate that the design can perform 40 multiply and adds in 40 µs using a single synapse.

5.7.8 Results from test chips

The floor plan for the chip design described in the previous chapter is shown in Figure 5.23. The chip design used 78 pins on a silicon die of 49.60 mm^2, although only 25 mm^2 were required for the design (the design was forced to be pad ring limited in order to fit the only suitable package available). The chip was designed with testability in mind, and therefore is in two distinct parts. In the centre is the main synapse array of eight neurons, each with 16 synapses multiplexed in two banks of eight. To the right of this is a small test

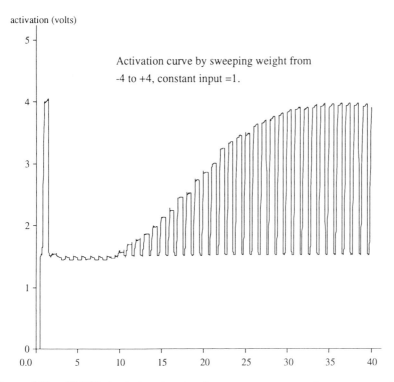

Figure 5.21. *SPICE simulation showing change in activation with synaptic weight – per-pulse system.*

section which contains two neurons each with two synapses. One of these neurons is a complete test neuron, the other is split with the summation op-amp removed so that the post-synaptic current can be routed to the pad ring whilst a separate input from the pad ring returns the summation voltage. Reference points are taken from the complete test neuron, from the synaptic weight capacitor, from the output of the op-amp sample-and-hold, from the ramp generating dummy neuron and from the output pulse streams. Other test pins on the chip monitor the output of the input sample-and-hold, the on-chip reference voltages (so that they could be externally overridden if necessary) and the master clock as primary power up check for the ICs.

To compare chip performance with results from the SPICE simulations, an interface board was built to control two digital to analogue converters (DACs) and read back voltages from the test chip. The test board provides the reference and test signals required to operate the chip, whilst the interface board allows two DACs to be set from a computer via RS232 serial lines, and can also read back voltages with an 8-bit analogue to digital converter (ADC). Using this test rig, the following results were obtained from the test chips.

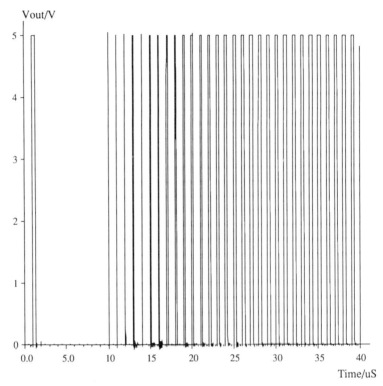

Figure 5.22. *SPICE simulation showing neuron switching ON as synaptic weight is swept – per-pulse system.*

5.7.9 Synapse linearity

The first test performed was to investigate the weight voltage/synaptic current relationship for the synapses. In Figure 5.24, the output of an external op-amp, configured as a current to voltage converter, is shown for the case when the weight voltage is swept through the voltage range of 1.0–5.0 V. The op-amp had a 91 kΩ feedback resistor and its non-inverting input was set at 2.5 V; thus a 1 V swing from 2.5 V represents a current of 10.1 μA. The simulation shows that the transistor gain was somewhat higher than expected (thin gate oxide), the maximum current being about 15 μA rather than 8 μA as designed. This figure also demonstrates the weakness of the SPICE models as the gate/source voltage approaches the threshold voltage V_T of the MOS transistors (sub-threshold or weak inversion region).

5.7.10 Input sample and hold

To test the pulse width sample-and-hold, the input bank select signal was driven by a 250 kHz clock, derived from, and synchronous to, the master

PER-PULSE COMPUTATION

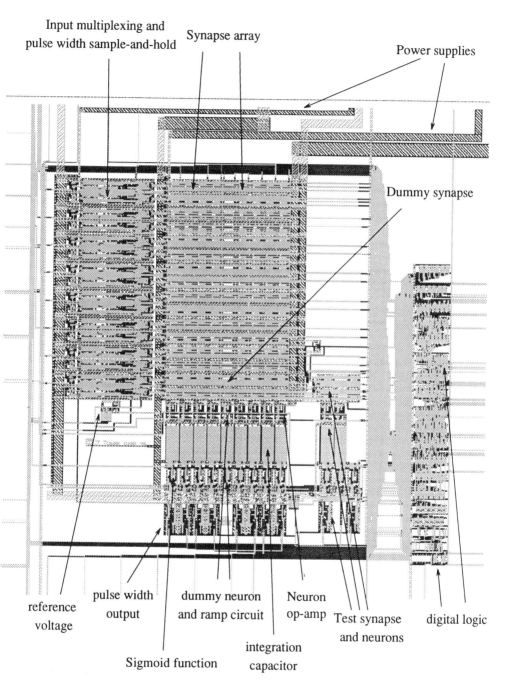

Figure 5.23. *Floorplan for the per-pulse chip.*

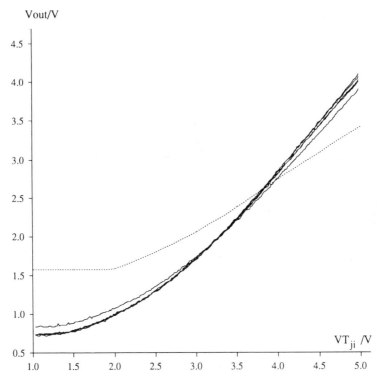

Figure 5.24. *Synapse (non) linearity for swept synaptic weight (per-pulse system).*

clock. Figure 5.25 shows the signal recorded from an oscilloscope when a 200 ns pulse is applied to the test chip with the master clock running at 1 MHz. The lower trace initially shows the input to the chip. During the period of the pass clock, the bank select signal goes low (upper trace) and the second pulse on the lower trace is generated from inside the chip by the pulse-width sample-and-hold circuit. For this (or larger) pulse widths, the sample-and-hold is very accurate on all chips tested, the difference in width between input and regenerated pulses not being measurable on a 50 MHz oscilloscope. As the input pulse width is made shorter, the observed error increases. Figure 5.26 shows a short (25 ns) input pulse and its regenerated pulse. In this case, the regenerated pulse is significantly shorter (≈ 20 ns) than the input pulse, corresponding to a 20% error, which represents the worst case, as this error was found to vary from 2–20% between chips (for an input pulse of 25 ns). The problems associated with short pulses are probably in part caused by the difficulties in ensuring that clock skew does not affect the effective length of the pulse and predominantly by the inaccuracies caused by having to operate the comparator at the limit of its common mode voltage range.

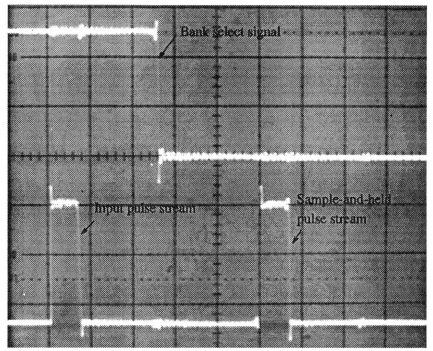

Figure 5.25. *Regeneration of a 200 ns pulse (per-pulse system).*

5.7.11 Sigmoid transfer function

A simple first-order low-pass filter (with a cut-off frequency of 6 kHz) can be used to measure the DC value of the output pulse stream. This is an indirect measure of pulse width and hence of the output of the sigmoid transfer function circuit. This low-pass filter was used to process the output of the sigmoid circuit as the linear summation value ($\Sigma T_{ij} V_j$) was swept across its range by varying the synaptic weights. The output of the sigmoid circuit is a short ramp which has a gradient the value of which depends on the linear sum for the period of the reset signal; it is then held at a constant value for the active period and finally returns to the value of V_{one} in the pass period (Figure 5.28). The constant value during the active period represents the value of the sigmoidal activity, but as the waveform is of fixed length, the low-pass filtered DC value provides a direct measure of this value. Figure 5.27 shows the low-pass filtered voltage generated as the weight voltage for the two test synapses is swept across the linear working range. The graph shows the two ICs which represented the limits of the variation between all the chips that were tested.

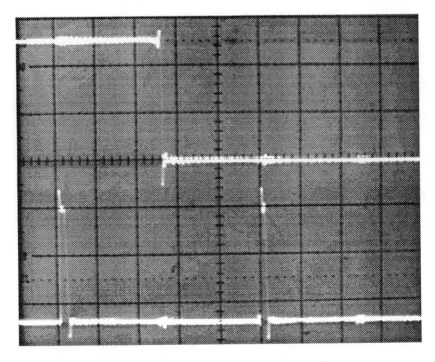

Figure 5.26. *Regeneration of a 25 ns pulse (per-pulse system).*

5.7.12 Output pulse stream generation

The relationship between sigmoid activation, sawtooth ramp and output pulse stream for two different sums is shown in Figures 5.28 and 5.29. These diagrams demonstrate the key parts of the design functioning together and show the reproducibility of the output for the two-synapse test neuron. From these diagrams, the trailing edge of the output pulse can be seen to correspond exactly to the point at which the ramp signal becomes equal to the sigmoid activation value. These signals were seen to be correctly produced and the chip to continue to function as the master clock was taken to frequencies as high as 6 MHz, although, at such high speeds the pulse widths are so short that meaningful measurement is not possible and could easily be swamped by clock skews across the chip.

5.7.13 Weight precision

The issue of the precision with which an individual synaptic weight can be set (to show a measurable change in the output of the post-synaptic neuron) is very important in the context of implementing suitable hardware learning strategies, as will be discussed in Chapter 7. Figures 5.30 and 5.31 demon-

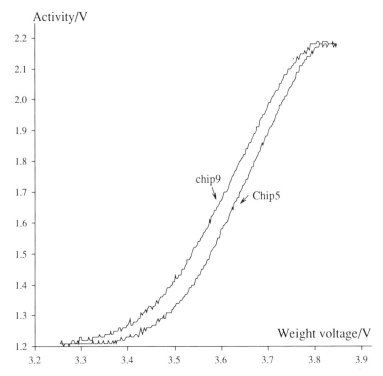

Figure 5.27. *Variation of activation with synaptic weight (per-pulse chip results).*

strate the change in output pulse width as the synaptic weight is changed by the least significant bit of the 8-bit DAC used to set the weights (in this case representing ±4 mV). The horizontal scale is 10 ns per division and the oscilloscope was triggered from the master clock. Figure 5.30 shows the trailing edge of the output square wave with the oscilloscope adjusted so that the mid point crosses at the (0, 0) point of the oscilloscope X-Y coordinates (point A). In Figure 5.31, the DAC value is reduced by one bit with the oscilloscope setting unaltered and the output pulse can be seen to be just over 4 ns shorter. These traces show that even with such resolution on the time axis, the output pulse width has an almost undetectable amount of noise superimposed on it. The jitter on the zero crossing of the falling edge is much less than the effect of changing the weight by one LSB. From this, we conclude that the individual weights can certainly be set to 1 part in 256 (since this leads to a measurable change in the output) and probably even to 1 part in 1000. Although it would be impossible to expect that all the synapses across a chip could be set with this precision relative to each other, this sort of result corresponds to the 1000 to 1 dynamic range (per individual synapse) observed by other groups with CMOS multipliers implemented in analogue VLSI [71].

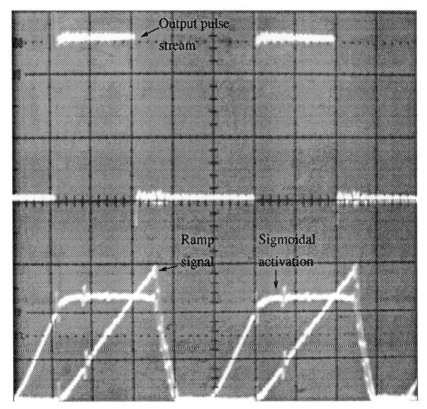

Figure 5.28. *Signals associated with the per-pulse neuron – from actual silicon.*

5.7.14 Weight update

The hold times for the synapse weight capacitor have also been measured; leakage is approximately 1.2 mV per second, which indicates that more than 10,000 synapses could easily be fabricated on one chip before the speed at which they needed to be refreshed became an issue.

5.7.15 Per-pulse computation – summary

Although the chip fabrication process is a *digital* process without true capacitors, novel pulse-stream neural network circuitry, optimized for the implementation of fast multi-layer perceptrons, has been designed and tested. Unlike many other analogue designs, it makes use of both parallel computation and fast analogue processing throughout. To capture the advantages of massively parallel computation, chips with as many as 10,000 synapses need to be fabricated. There is no reason why our test chips cannot be scaled up to make this possible; with the master clock

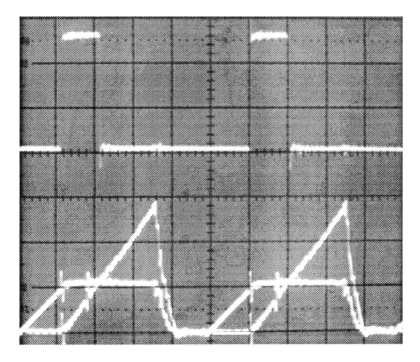

Figure 5.29. *Signals associated with per-pulse neuron – from actual silicon.*

frequency set at a frequency of 1 MHz, 10^{10} operations per second would be performed by a single chip. This should *not* be converted to a figure in MFLOPS as with digital computers, since the operations are low-precision analogue multiply-and-accumulate computations; nevertheless, the figure does show what can be achieved with pulse stream technology when speed of computation is made the most important criterion.

5.8 EPSILON – the chosen neuron/synapse cells, and results

This section describes the Edinburgh Pulse-Stream Implementation of a Learning Oriented Network (EPSILON) chip in detail. The main design criteria were as follows:

1. That it be large enough to be of use in practical problems.
2. It should be capable of implementing networks of arbitrary size and architecture.
3. It must be able to act as both a 'slave' accelerator to a conventional computer, *and* as an 'autonomous' processor.

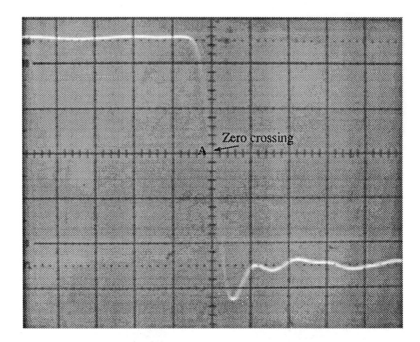

Figure 5.30. *Output pulse from per-pulse chip.*

These constraints resulted in a chip which could realize only a single layer of synaptic weights, but which could be *cascaded* to form large, useful networks for solving real-world problems.

Many of the circuits employed on EPSILON have already been presented, and described in Sections 5.1–5.5 – we will simply summarize them here to make clear what EPSILON actually comprises. Finally, results from a vowel recognition application are presented to illustrate the performance of EPSILON when applied to real tasks, configured as a multi-layer perceptron – the most widely-used neural network architecture.

5.8.1 The EPSILON design

Here, the circuits used are described – we simply draw out any changes that were made in including them in EPSILON, and present characterization results. In accordance with the demerits mentioned in Sections 5.1–5.5, all circuits were designed to be tolerant to noise and process variations, to *cause* as little noise as possible themselves, and to be easy to 'set up' in practice. Naturally, we are not 100% happy with the results, but they are nonetheless impressive. Finally, the specification of the EPSILON chip is presented, before moving on to demonstrate its capabilities as a network.

EPSILON – THE CHOSEN NEURON/SYNAPSE CELLS, AND RESULTS 87

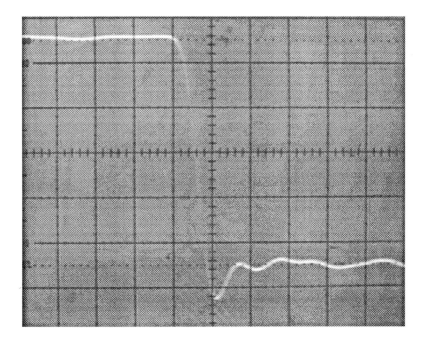

Figure 5.31. *Output pulse from per pulse chip – with a 1-bit change from Figure 5.30.*

5.8.2 Synapse

The synapse design was based on the standard transconductance multiplier circuit. Results from characterization tests of the synapse are presented in Figure 5.32, which shows output state against input state, for a variety of different synaptic weight voltages. As seen from the Figure 5.32, the linearity

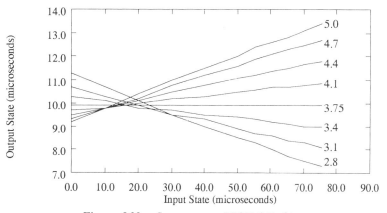

Figure 5.32. *Synapse test, EPSILON chip.*

of the synapses, with respect to input state, is very high. The variation of synapse response with synaptic weight voltage is also fairly uniform. The graphs depict mean performance over all the synaptic columns in all the chips tested. The associated standard deviations were more or less constant, representing a variation of approximately ± 300 ns in the values of the output pulse widths. The effects of across- and between-chip process mismatches would therefore seem to be well contained by the circuit design. The 'zero point' in the synaptic weight range was set at 3.75 V and, as can be seen from Figure 5.32, each graph shows an offset problem when the input neural state is zero. This was due to an imbalance in the operating conditions of the transistors in the synapse, induced by the non-ideal nature of the power supplies (i.e. the non-zero sheet resistance of the power supply tracks), resulting in an offset in the input voltage to the post-synaptic integrator. This problem is easily obviated in practice, by employing three synapses per column to cancel the offset.

5.8.3 Neurons

To reflect the diversity of neural network forms, and possible applications, two different neuron designs were included on the EPSILON chip. The first, the *synchronous* pulse width modulation neuron of Section 5.5 was designed with vision applications in mind. This circuit could guarantee network computation times, thereby eliminating the data dependency inherent in pulse frequency systems. The second neuron design based on that in Section 5.4 used *asynchronous* pulse frequency modulation; the asynchronous nature of these circuits is advantageous in feedback and recurrent neural architectures, where temporal characteristics are important. As with the synapse, both circuits were designed to minimize transient noise injection, and to be tolerant of process variations.

Pulse width modulation

The operation of the pulse-width modulation neuron is illustrated again in Figure 5.33, indicating how a sigmoid characteristic is achieved. Note that the sigmoid signals are 'on their sides' – this is because the input (or independent variable) is on the vertical axis rather than the horizontal axis, as would normally be expected.

Figure 5.34 shows plots of output state (measured as a percentage of a maximum possible 20 μs pulse) versus input activity voltage, for five different sigmoid temperatures, averaged over all the neurons on one chip. As can be seen, the fidelity of the sigmoids is extremely high, and it should be noted that all the curves are symmetrical about their midpoints – something which is difficult to achieve using standard analogue circuits.

EPSILON – THE CHOSEN NEURON/SYNAPSE CELLS, AND RESULTS

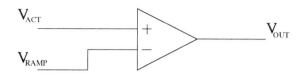

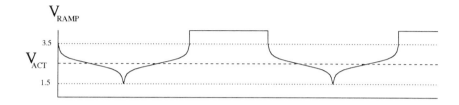

Figure 5.33. *PWM neuron with sigmoidal characteristic.*

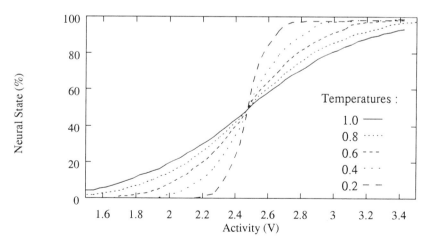

Figure 5.34. *Operation of PWM neuron with varying sigmoid 'temperatures' – from EPSILON chip.*

Pulse frequency modulation

The second neuron design which was included in EPSILON used pulse frequency encoding of the neural state (Section 5.4). Although subject to data dependent calculation times, its wholly asynchronous nature makes it ideal for neural network architectures which embody temporal characteristics, i.e. feedback networks, and recurrent networks. The characterization results for the VCO are presented in Figure 5.35, which shows plots of output percentage duty cycle *versus* input differential voltage, for different values of sigmoid gain. Note that the curves are fair approximations to sigmoids, although, in contrast with the pulse width modulation neuron, they are *not* symmetrical about their mid-points. It can also be seen that the range of possible sigmoid gains is smaller than the range available with the PWM system, although this is not a crucial factor in many applications.

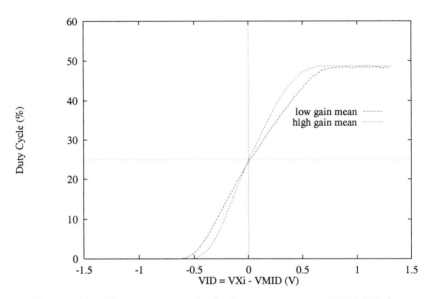

Figure 5.35. *Characterization of pulse frequency neuron – EPSILON chip.*

5.8.4 EPSILON specification

The circuits described in the previous section were combined to form the EPSILON chip. This was subsequently fabricated by European Silicon Structures (ES2) using their ECPD15 (i.e. 1.5 μm, double metal, single polysilicon CMOS). As already stated, each chip was capable of implementing a single layer of synaptic connections, and could accept inputs as either analogue voltages (for direct interface to sensors) or as pulses (for communication with other chips, and with digital systems). The specification is given in Table 5.1, with similar figures for the Intel ETANN chip included for comparison.

Table 5.1 EPSILON and ETANN specifications

ETANN vs. EPSILON – Comparison	
80170NX (ETANN)	30120PI (EPSILON)
Floating gate technology	Standard CMOS process
64 variable gain neurons	30 variable gain neurons
128 inputs	120 inputs
10,240 synapses	3,600 synapses
$2 \times (80 \times 64)$ arrays	120×30 array
Inputs: analogue	Inputs: analogue, PS and PWM
Outputs: analogue	Outputs: PS and PWM
3 µs per layer	10 µs per layer (PWM)
2B connections/s/chip	0.36B connections/s/chip
Non-volatile weight storage	Capacitive weight storage
>100 µs per weight update	1 µs per weight update
Multiplication non-linear at *extrema*	Linear multiplication over weight range

5.8.5 Application – vowel classification

After the device characterization experiments had been completed, EPSILON was used to implement a multi-layer perceptron (MLP) for speech data classification. The MLP had 54 inputs, 27 hidden units and 11 outputs, and the task was to classify 11 different vowel sounds spoken by each of 33 speakers. The input vectors were formed by the analogue outputs of 54 band-pass filters. This particular database will be returned to in Chapter 7, in the context of on-chip learning. While not perhaps a truly **difficult** neural task (a deliberately difficult task will also be described in Chapter 7) it does represent a large database of analogue data, and the 1-out-of-11 output coding allows classification to be made on the basis of *a posteriori* probabilities. It is thus a good problem on which to demonstrate a large neural chip.

The MLP was initially trained on a SPARC station, using a subset of 22 patterns. Learning (using the Virtual Targets algorithm – described in the final chapter of this book – with 0% noise [72]) proceeded until the maximum bit error in the output vector was ≤ 0.3, at which point the weight set was downloaded to EPSILON. Training was then restarted under the same regime as before (using the same 'stop' criterion), although this time EPSILON was used to evaluate the 'forward pass' phases of the network. Figure 5.36 shows the evolution of mean square error with number of epochs during this period; at the end of training, EPSILON could correctly identify all 22 training patterns.

Subsequent to this, 176 'unseen' test patterns were presented to the EPSILON network, with the result that 65.34% of these vectors were correctly classified. This compared very favourably with similar

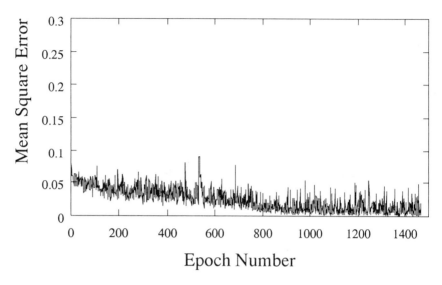

Figure 5.36. *Training the EPSILON chip – vowel classification data.*

generalization experiments which were carried out on a SPARC: in this case, the best result obtained was 67.61%.

5.9 Conclusions

We have designed analogue circuits for neural network computation in a process invariant and cascadable manner. These circuits have, by and large, not only worked, but worked well. As a result, we have been able to design a complex demonstrator chip, from which preliminary results are very exciting.

More generally speaking, however, there are a number of 'practical' tips which we feel would be of use/interest to the aspiring neural VLSI designer. They apply, of course, to all analogue VLSI design:

- Think long and hard about how the vagaries of an analogue process will impinge on your circuits and ideas **before** you set mouse to layout, and design your circuits to alleviate these problems as much as possible. Don't hesitate to use any 'clever' external tricks to help attain your desired end.
- Simulate, simulate and simulate again. The design should be tested over all combinations of process extremes to maximize its chances of functioning as desired.
- Where possible, try all new designs on small test chips before committing yourself to a major chip.
- Assign as many output pins as possible to key signals, references and set-points. In the event of a failure, you will want to extract as much

information as you can from the chip, so that the fault can be rectified or circumvented in future design iterations. SPICE can be useful here as a direct comparison with your silicon. Remember, however, that making as many internal points available for testing as possible will conflict with your desire to make the test chip as small and inexpensive as possible.
- Perhaps more realistically, even if you think that your chances of obtaining a working synapse array **first time** are high enough that the test chip is configured as a network, **place a single synapse in the corner, with everything pinned out**. Then you should be able to figure out what went wrong when the array fails to perform as expected.
- Finally, and perhaps a little obvious – keep good documentation. Depending on your silicon source, it can be a matter of months between design submission and the return of chips from the foundry. In the meantime, you've certainly forgotten which pin was which.

Despite the many mishaps that led to the above list of 'things-to-watch-out-for', EPSILON – a large analogue VLSI neural chip, composed of process tolerant circuits, with useful characteristics – has been fabricated. Although not a *self*-learning device, it has been proved that EPSILON will support learning, and can be applied successfully to real-world problems. Indeed, when correctly trained, the performance of the EPSILON chip has been shown to be comparable with that of software simulations on a SPARC station.

Acknowledgements

The authors of this chapter would like to thank the Science and Engineering Research Council for their continued funding of this work. In addition, Stephen Churcher and Donald Baxter are grateful to British Aerospace plc and Thorn-EMI CRL respectively for sponsorship and technical support during the course of their PhDs.

6

Application examples

6.1 Introduction

As we made clear in the introductory chapter to this book, one of the principal motivations for our work on pulse-stream VLSI neural networks was the desire to exploit the advantages of hardware parallelism to obtain real-time solutions to signal processing and pattern recognition problems. The first set of chips to be used in a real-time demonstrator were the switched-capacitor test chips described in Chapter 5. Ten prototype chips (each including 12 neurons and 144 synapses) were fabricated; of these, eight were found to be fully functional and two of those were used in our first demonstrator, the real-time recognition of isolated words.

6.2 Real-time speech recognition

With this application, the input data to each of the two test chips is a succession of 12-dimensional vectors, each vector $\mathbf{u}_i$ representing the energy of the speech waveform in the frequency domain during consecutive 20 ms intervals.[1] The speech output from a microphone is amplified and fed to a bank of 12 analogue band-pass filters whose centre frequencies are chosen to give uniform coverage (on a logarithmic scale) of the speech spectrum between 300 Hz and 4 kHz. The output of each band-pass filter is rectified and low-pass filtered to give an *analogue* measure of the energy in that band. Templates of the words in the vocabulary are created by storing in memory the set of consecutive 12-dimensional 'energy vectors' generated every 20 ms by the spoken utterance; thus, for a one-second word, there would be 50 $\mathbf{u}_i$ vectors to be stored, each consisting of 12 analogue values.

The recognition of words stored in the vocabulary was carried out by using the two switched-capacitor chips configured as *minimum distance*

---

[1] The spectral properties of speech are known to be *quasi-stationary* over such a time interval, and thus the spectrum of the speech waveform need not be computed more often than every 20 ms or so.

classifiers. As its name implies, a minimum distance classifier is a neural network which classifies input patterns on the basis of their distance from a set of exemplar patterns, or templates, stored in memory. The distance metric is usually chosen to be the *Euclidean* distance. As shown below, a set of M perceptrons (see Chapter 1) operating in parallel can be used to identify which of the M templates is closest to the pattern presented at the input of the network. The use of two chips means that the word-recognition experiments can be carried out on a vocabulary of 24 words (alternatively, a vocabulary of 12 words could have been used, with two versions of each word stored separately in order to improve recognition performance). The recognition takes place in real time, with a new 12-dimensional vector $\mathbf{x}$ being generated every 20 ms from the processed outputs of the bank of band-pass filters. The corresponding 12-dimensional vectors $\mathbf{u}_i$ stored in memory (i.e. the templates) are also read out from memory consecutively, every 20 ms. (We are assuming, with this simple demonstrator, that each utterance of the same word is spoken at approximately the same rate.) Chapter 3 showed that analogue memory technology is not yet a viable concept and hence digital memory had to be used to store the templates. The sets of 50 consecutive 12-dimensional vectors making up each exemplar word were digitized to 8-bit accuracy and stored in digital RAM (i.e. 600 bytes per template). Once retrieved from memory, the $\mathbf{u}_i$ vectors are applied, via a D/A converter, to the chips as the set of T_{ij} weights for that frame,[2] for reasons which are explained immediately below. The chips compute the Euclidean distances between $\mathbf{x}$ and each of the 24 $\mathbf{u}_i$s (in parallel, for each 20 ms frame) as follows:

$$\|\mathbf{x} - \mathbf{u}_i\|^2 = \|\mathbf{x}\|^2 - 2\mathbf{u}_i^T\mathbf{x} + \|\mathbf{u}_i\|^2 \tag{6.1}$$

The first term in the above equation is the same for all i and can be ignored. We can therefore write:

$$g_i(\mathbf{x}) = -\tfrac{1}{2}(-2\mathbf{w}_i^T\mathbf{x} + \mathbf{u}_i^2) = \mathbf{w}_i^T\mathbf{x} + w_{i0} \tag{6.2}$$

where $g_i(\mathbf{x})$ is a linear discriminant function, $\mathbf{w}_i = \mathbf{u}_i$ and $w_{i0} = -\tfrac{1}{2}\mathbf{u}_i^2$. If we let $\mathbf{w}_i = \{T_{ij}\}$ and $\mathbf{x} = \{V_j\}$, then we can write:

$$g_i(\mathbf{x}) = \sum_{j=1}^{j=12} T_{ij}V_j + w_{i0} \tag{6.3}$$

Thus, for each 20 ms frame, the $\mathbf{u}_i$ vectors from each template are used as the set of synaptic weights for the network and the neurons g_i can be thought of as processors which compute the distances between $\mathbf{x}$ and each $\mathbf{u}_i$. This is the well-known implementation of the minimum distance classifier on a neural network architecture. Because of the minus sign in equation (6.2), the pattern $\mathbf{u}_i$ to which $\mathbf{x}$ is closest is indicated by the neuron $g_i(\mathbf{x})$

---

[2] Each 20 ms interval is known as a 'frame'.

which has the *maximum* output. The evaluation of equation (6.3) takes place in parallel for all 24 templates, once every 20 ms, and the result is stored on each neuron's output capacitor. Moreover, the use of a capacitor to store the result means that the results from successive frames can be *accumulated* and the unknown word recognized as soon as it has been spoken by identifying the 'most active' neuron, i.e. the one with the highest voltage on its output capacitor at the end of the utterance.

Our simple experiments showed that, even without any attempts at optimization of the templates, recognition rates between 80% and 90% were possible if the vocabulary was chosen to include only words of similar duration. (A less restricted vocabulary would have required some form of time normalisation.) This figure, of course, depends both on the vocabulary and the speaker chosen, and the whole exercise was only ever meant to be a first demonstration of the real-time capabilities of the pulse-stream VLSI hardware.

6.3 Applications of neural VLSI

Of course, the real-time recognition of isolated words can also be achieved with a DSP chip running a neural network algorithm. There is no need to use an analogue VLSI chip. The following question must, therefore, be asked at this stage: what can be done with massively parallel analogue technology which could not be done with a digital computer (or DSP chip) running a neural network algorithm? We believe that there are at least three different areas in which parallel analogue VLSI has, or will have, significant advantages:

- Dedicated systems which require parallelism for *real time* processing and VLSI for portability and low power consumption
- Hardware co-processors
- Embedded neural systems with on-chip, on-line learning.

6.4 Applications of neural VLSI – dedicated systems

There are a few applications for which the need for a massively parallel VLSI implementation is clear because real-time processing must be achieved within constraints of power, cost or portability (or any combination of these). Such applications are typically found in low-level vision or mobile robotics or they may be found within the realm of space technology, for example. In this section, we concentrate on a real world problem which has not been *fully* solved by traditional Artificial Intelligence (AI) methods: the *real time* control of an autonomous mobile robot navigating in a structured environment. The most common application of this is to be found in automated factories, where Autonomous Guided Vehicles (AGVs) are required to transport parts around the factory. The type of system currently in use has very limited perception capabilities and as a result can only perform very specific tasks in static environments. Typical of the current technology are the *wire-guided* systems in which the AGV follows a wire buried in the floor. If an obstacle appears in its path, the robot can only (at best!) stop and wait for the obstacle to move or to be moved. More recently, free-ranging systems have begun to appear in more advanced factories which are populated with artificial *beacons* (sometimes also known as landmarks) such as retro reflecting bar-codes. However, the robot must always be within sight of at least two beacons (preferably more) and it can only cope with obstacles whose position is known *a priori*. Currently under development is the *next* generation of AGVs, robots capable of navigating in real time in *unmodified* environments (i.e. with no beacons) in which unexpected obstacles may appear from time to time, for example other vehicles and packages. Even when partial solutions from AI have been proposed and implemented in an attempt to tackle this problem, these have required vast computational resources, usually remote from the AGV and linked to it via an umbilical cord. The neural network modules described in this section can provide the basis for a navigation system capable of real time operation (because of the massive parallelism) for a truly autonomous robot (the VLSI implementation keeping the power and weight requirements to a minimum).

The key requirement of an AGV navigation system is for a *path planning* module which can not only calculate a path from the start to the goal but also new paths whenever unexpected obstacles are encountered. A *localization* module is also essential to determine, at any time, the robot's position within the environment for the following reasons:

- Knowledge of current position is required to re-plan a new route after an obstacle avoidance manoeuvre has been executed.
- Accumulated errors in robot motion (due, for example, to the unevenness of the floor) will cause the robot gradually to deviate from its initially planned route.
- In certain cases (shortage of battery power, for example), the robot might have to return to specific locations within its working environment.

Finally, as already hinted above, an *obstacle detection/avoidance* module is required so that evasive action can be taken when a moving object is approaching the robot at such a speed and angle that a collision will almost certainly occur. The reaction time of this module must be fast enough to prevent the robot from colliding with the fastest moving obstacles within the environment. In the overall scheme shown in Figure 6.1, the VLSI neural network modules perform *low-level processing in real time* which is then decoupled from the higher level processing of this data carried out by a central controller. It is our opinion that such a hybrid system is the best

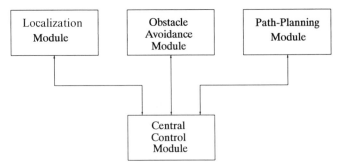

Figure 6.1. *Mobile robot navigation system.*

means of exploiting the computational potential of artificial neural networks in this particular type of application.

Each neural network module is now briefly described below. The robot's workspace is modelled as a 2D environment, for which it is assumed that the boundaries and the locations of any permanent obstacles (e.g. pillars) are known *a priori*. This environment is discretized, for path planning and localisation purposes, using a grid of hexagonal or rectangular cells.

6.4.1 Path planning

The path planning module is a *resistive grid* of hexagonal or rectangular cells used to compute a collision-free path in the robot's workspace from its initial (or present) position (P) to the goal (G). Each node is connected to its nearest neighbour (four for a square grid, six in the hexagonal case) by a resistor whose value is, say, R_0 unless it is part of a region of the grid corresponding to an obstacle, in which case the resistance is infinite (R_∞). As soon as an external constant current source is connected between P and G, the resistive network settles into the state of least power dissipation. The path is then computed from *local* voltage measurements: at each node, the next move is identified by measuring the voltage drop ΔV_n between that node and its nearest neighbours ($n=6$ for a hexagonal grid) and then selecting the node corresponding to $(\Delta V_n)_{\max}$ (steepest gradient descent in V). A discussion of the problems which can occur with this simple strategy (for example, how to cope with local minima) is beyond the scope of this book and the interested reader is referred instead to the papers by Tarassenko *et al.* [73, 74].

Resistive grids cannot be said to be neural networks in the conventional sense. However, they are extensively used by Mead[3] and described as such in his book [23]. In any case, resistive grids, like neural networks, undoubtedly perform parallel analogue computation and they have the same

---

[3] For example, in his silicon retina [30] or for computing optical flow [75].

advantages, in terms of speed and fault-tolerance, as any massively parallel hardware realisation of neural networks.

6.4.2 Localization

The grid representation used for the path planner is also employed for localization purposes, in which case localization becomes a matter of identifying the *nearest node* in the grid at any time during navigation. As with the speech recognition problem described in Section 6.2, this task can be performed in real time using a neural network configured as a minimum distance classifier.

The room environment is first of all 'learnt' during a training phase by recording a 360° range scan at every node in the grid using a rotating time-of-flight infra-red scanner mounted on the robot. The use of light, rather than ultrasound as with most current AGV scanners, makes real-time localization possible and avoids the problem of specular reflections. During navigation, the nearest node on the grid is identified by comparing the most recently acquired range scan with each of the patterns stored during the training phase. The number of points over which the comparison is carried out depends on the number of synaptic weights available on the chip (see below). If the orientation of the robot is unknown, then not only must a scan be stored for every node on the grid but also for every possible orientation (say every 3°). Normally, an on-board compass can provide an approximate indication of the robot's orientation and so the storage requirements are drastically reduced. Although the data is generated *serially* by the infra-red scanner as it rotates, real time localization with the minimum distance classifier requires a comparison with many different scans in parallel.

At the time of our initial experiments, the only chips available to us were the switched-capacitor test chips with 12 synapses per neuron which meant that the comparison could only be performed on 12 points per scan (i.e. one every 30°). This quantization is too coarse to give adequate results in a real environment and so the test chips were tested instead on a manufactured *test problem* using an idealised room and simulated range data. For the purposes of simulation, the model room of Figure 6.2 was constructed and mapped out with 24 grid points (as with the speech recognition problem, two test chips were used). The room environment was 'learnt' by using the simulation programme to generate the 12 range readings which an ideal range scanner would have given every 30° (the effect of varying orientation was neglected in this simple pilot study – the 12 range values for both the stored scans and the scans generated during the simulated navigation phase were ordered in the same sequence, always starting with the range value obtained at an angle of 0° with respect to the room x-axis). The 24 sets of analogue range values were then normalized

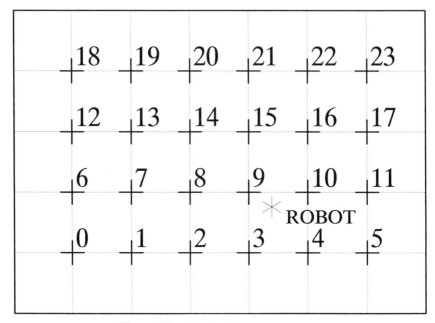

Figure 6.2. *Simulated test environment.*

to be of equal energy[4] and became the T_{ij} weights for the two chips. During the navigation phase, the robot was 'taken' on a random path within the room and scans acquired along this path were fed as the input V_j values to the two chips which then computed in parallel, for all grid points, the distance between the most recent scan and all the stored scans by evaluating the 24 $T_{ij}V_j$ scalar products. The robot's approximate position (i.e. the nearest grid point) was then obtained, as in the speech recognition problem, by identifying the neuron with the highest activity.

As an example, consider the case when the robot is at the point marked X on Figure 6.2. The scan recorded at this position is fed as a set of 12 input V_j values to each of the two chips. The results obtained with the chips were compared with a full simulation of the same matching procedure carried out on a SUN 3/80 workstation. In each case, the T_{ij} weights were again quantized to the 8-bit accuracy of the D/A converter used to refresh the chip weights. The $\Sigma T_{ij}V_j$ values obtained for each grid point in both cases are shown in Figure 6.3; note the limited range of the vertical axis. The figure shows that the scalar products evaluated by the two VLSI chips are within 1.2% of those computed on the SUN workstation.[5] Most importantly, this means that grid point 9 is correctly

---

[4] This means that the w_{i0} term in equations (6.2) and (6.3) can be ignored.
[5] This might be taken to correspond to a precision of 6–7 bits, except that assigning a 'digital precision' to analogue computation can be a misleading exercise – see next chapter.

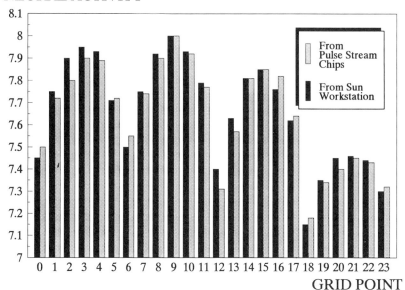

Figure 6.3. *Comparison of VLSI chips and SUN workstation on localization test problem.*

identified by the switched-capacitor chips as being the closest to the robot.

6.4.3 Obstacle detection/avoidance

This module will rely on optical flow information derived from a number of *fixed* optical sensors mounted on the robot platform. Ambient light reflected from nearby objects will be focused onto a pair of gratings at right angles to each other, before being detected by a photodiode array. The optical flow field contains information on the relative velocities of nearby objects, such as moving obstacles. We plan to use reinforcement learning [76] to form associations between feature vectors derived from the velocity vectors (the sensory input patterns) and predictions of future outcomes. Actions leading to the 'desirable' outcomes (principally the avoidance of moving obstacles) will be 'rewarded', whereas those actions which lead to eventual collisions will be 'punished' during the learning phase, so that the sensory-motor associative network of Figure 6.4 is gradually evolved. The work on this module has, so far, concentrated on the design of the analogue visual motion sensors and on the simulation of reinforcement learning but, again, the analogue VLSI technology described in this book will be ideal to process the

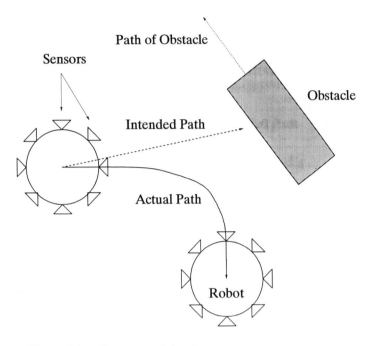

Figure 6.4. *Illustration of the obstacle avoidance module in use.*

high-bandwidth data produced by the 20 or so analogue motion sensors distributed around the periphery of the robot.

6.4.4 Conclusion

This brief description of our VLSI robot navigation system (which remains to be fully implemented) has been included to demonstrate that, for some problems, the development of parallel analogue VLSI is the key requirement to meet the goals of the project. In this case, the massive parallelism, small size and low power consumption of a neural VLSI solution are required for the successful implementation of a truly autonomous robot, with all the (battery-powered) sensors and control hardware mounted on the platform, and capable of *real time* navigation in factory-like environments.

6.5 Applications of neural VLSI – hardware co-processors

If a software programme is examined closely, it is often found that the majority of the execution time is spent on a few time-consuming instructions. If these were implemented on a hardware co-processor instead, then it may be possible to obtain an increase in processing speed of several orders of magnitude (the same effect was seen when floating-point co-processors were first introduced). We have been exploring how certain aspects of the parallel analogue technology described in this book could be used in hardware co-processors for the extremely fast generation of an approximate solution to problems such as equation solving. The accuracy of solution would then be improved by the digital computer in a few iterations, rather than the much larger number that would have been required if the computer had started from a random guess, as is often the case.

6.6 Applications of neural VLSI – embedded neural systems

Although we hope to have demonstrated to the reader by now that the technology for implementing neural networks in analogue VLSI does exist, stand-alone neural chips which are *capable of learning* without the support of a host processor have not yet been built. As with the robot localization module described in Section 6.4.2, the synaptic weights which are loaded into the VLSI chip are invariably derived during a *prior* training phase which takes place *off-chip* on a conventional computer. Thus the requirement for a digital computer, at least during the learning phase, has not been eliminated. We believe that the development of hardware learning for analogue VLSI is now such a key issue that we have devoted most of the last chapter of this book to it.

6.7 Conclusion

This short chapter has been biassed towards the kind of applications of interest to the authors of this book. Nothing has been included, for example, on character recognition for which impressive hardware demonstrators have been built [77]. Our application work has so far been limited to speech recognition and mobile robotics because most of our effort has been concentrated on designing VLSI circuits. The simple experiments described in this chapter, using single-layer perceptrons configured as minimum-distance classifiers, are not intended to demonstrate the power of analogue VLSI technology, but rather its applicability to a range of engineering problems. More sophisticated multi-layer networks are required before the power of the technology can be fully appreciated. We are at present working towards the incorporation of such networks both in diagnostic instrumentation embedded in experiments and in fully decentralized intelligent sensor systems.

7

The future

7.1 Introduction

If we are to build learning machines in analogue VLSI using the neural paradigm, then we have to learn how to design parallel analogue hardware with the ability to learn not only 'on-line' but also 'on-chip'. This means that we have to come up with learning strategies which allow the neural network to continue to *adapt* to changes in input data *beyond* the initial training phase; furthermore, we have to make sure that these learning strategies can be mapped directly into silicon with minimal external support circuitry.

If the twin goals of on-line and on-chip learning can be met, there is no doubt that adaptive neural machines could be used on a whole range of problems. The following two are offered simply as typical examples:

- adaptive mobile robot navigation, with an initial exploration phase during which the robot learns, from its sensory input data, the features of its working environment. In most cases, it is very likely that this environment will undergo some changes in the course of time, hence the requirement for adaptability beyond the original learning phase.
- handwritten character recognition, whereby a neural network, after initial training on a large database, may have to adapt to the handwriting of a new subject without having to re-train completely using the expanded database.

In both of these examples, hardware parallelism is required for real time performance, as software implementation of a suitable neural network algorithm on a serial computer would not be able to cope with the high bandwidth of the incoming data. There are two forms of on-line learning which can be envisaged: *incremental* and *sequential* learning. With incremental learning, synaptic weights are modified if the incoming input patterns are sufficiently novel with respect to those which are part of the database with which the network was initially trained. With sequential learning, the synaptic weights are modified with every new data sample, or input pattern, whilst at the same time the architecture evolves, albeit at a slower rate.

There is a significant amount of work being done in both of these areas at the moment [78], and indeed, some of our own research effort has recently been devoted to incremental learning [79].

It is premature, however, to come to any sort of conclusion about the outcome of such recent work or its eventual implications on hardware learning. For the time being, we have preferred to concentrate on a difficult enough but more established problem: on-chip learning with multi-layer networks. That is not to say that there are no applications within this more restricted field: one example of a possible application might be the real-time tracking of a solid object using a video camera. An operator would select the object to be tracked by the system and the neural network would then learn to recognize the object over a run of consecutive frames. (Although the problem of segmentation may seem daunting for such an application, it has been shown recently [80] that a multi-layer network working on raw pixel data with minimal pre-processing is capable of excellent digit recognition.) With on-chip learning, the operator could then select another object from the same, or a different scene, and the multi-layer network would then be able to learn this new task under operator control.[1]

The question which we are posing in our preliminary work on hardware learning is this: can the training of a multi-layer perceptron be implemented in hardware? Although multi-layer perceptrons are capable only of *batch* learning (the whole of the training database must be supplied during the learning phase), the rest of this chapter will show that a surprising amount can be learnt already from attempting to answer that question.

7.2 Hardware learning with multi-layer perceptrons

It is not clear from the existing literature whether the error back-propagation algorithm (with or without modifications) can be implemented in analogue technology. Most authors argue that the precision required is well beyond the reach of analogue technology – see, for example, [82]. A few others take the opposite view, arguing that learning is still possible when the back-propagation equations are mapped onto analogue hardware (for example [83, 84]).

It was shown in Chapter 1 that learning with multi-layer perceptrons is an iterative process, whereby an error function is minimized as successive input–output pairs are presented to the network. Gradient descent is used to minimize the mean square error E, the synaptic weights being adjusted

---

[1] In the context of this application, there is now the intriguing possibility that recently-developed CMOS sensor arrays [81] and our pulse-stream technology could be integrated on the same CMOS chip, giving rise to a trainable vision processor *on one chip*.

according to the following equation (see Chapter 1):

$$\Delta T_{kj} = -\eta \frac{\partial E}{\partial T_{kj}} = -\eta \frac{\partial E}{\partial x_k} \cdot \frac{\partial x_k}{\partial T_{kj}} = -\eta \delta_k V_j \qquad (7.1)$$

where η is a learning rate parameter, V_j is one of the inputs to neuron k, $\delta_k = \partial E/\partial x_k$, with $x_k = \Sigma T_{kj} V_j$ and $V_k = f(x_k)$. For an output neuron k, we have:

$$\delta_k = \frac{\partial E}{\partial V_k} \cdot \frac{\partial V_k}{\partial x_k} = V_k - D_k \cdot V_k (1 - V_k) \qquad (7.2)$$

where D_k is the desired output and V_k the actual output. If the neuron is an internal neuron j from the hidden layer, then

$$\delta_j = V_j (1 - V_j) \sum_k \delta_k T_{kj} \qquad (7.3)$$

where k is over *all neurons in the layer above the hidden layer*, i.e. the output layer for networks with a single hidden layer. Knowledge of all the δs in the next layer is therefore required to compute the weight update equation for a hidden layer neuron. Storing and routing that information in an analogue VLSI design would be a non-trivial problem.

The final difficulty with implementing back-propagation in analogue hardware arises because of the two-way flow of information, for *each* synaptic weight, imposed by the nature of the back-propagation algorithm. The synapse circuit must be capable of incrementing and decrementing the value of its weight, while passing signals forwards and error signals backwards. The design of a 'two-way' multiplier would inevitably impose further area and complexity overheads.

The evidence presented here shows that there are formidable obstacles to the implementation of the back-propagation algorithm in analogue VLSI. To develop more appropriate *hardware* learning strategies, we would suggest that there are two ways that the analogue VLSI designer can go:

- Extensive modification of error back-propagation until it is suitable for on-chip implementation (the 'top-down approach').
- Direct hardware learning strategies (the 'bottom-up approach').

We have begun an investigation of *both* of these approaches and our preliminary results are described in the next two sections of this chapter.

7.3 The top-down approach: virtual targets

This learning scheme introduces an explicit 'desired state', or target, for each of the hidden units, which is updated continuously and stored along with the synaptic weights. Although this means that a target state must be stored for each input pattern and hidden node, it simplifies and renders homogeneous the process of weight evolution for all neurons. Furthermore, since a target state is already stored for each output neuron, the scheme essentially removes the distinction during learning between hidden and output nodes. In effect, training a multi-layer perceptron with one hidden layer becomes equivalent to training two single-layer perceptrons.

The fundamental idea of adapting the internal representation as well as the weights is not new, but prior work in this area has not been optimized for hardware implementation. The key difference is that *simplicity* of implementation has been made the primary goal of the work described in this section to produce a system optimized for analogue VLSI.

7.3.1 'Virtual Targets' method – in an $I:J:K$ MLP network

The J hidden- and K output-layer neurons obey the usual equations (e.g. $V_k = f(x_k)$, where $x_k = \Sigma T_{kj} V_j$). Weights evolve in response to the presentation of a pattern p via perceptron-like equations similar to those used in back-propagation:

$$\frac{\delta T_{kj}}{\delta t} = \eta_{weights} V_{jp} o'_{kp} \varepsilon_{kp} \tag{7.4}$$

$$\frac{\delta T_{ji}}{\delta t} = \eta_{weights} V_{ip} o'_{jp} \varepsilon_{jp} \tag{7.5}$$

where, for instance, output layer errors are $\varepsilon_{kp} = \tilde{V}_{kp} - V_{kp}$, where $\{\tilde{V}_{kp}\}$ are the target states.

The terms V'_{kp} etc. represent the derivatives of the activation function $\delta V_{kp}/\delta x_{kp}$, which effectively discourage learning on weights which connect **to** neurons that are firmly OFF or ON. The terms V_{jp} and V_{ip} discourage learning on weights which connect **from** neurons that are firmly OFF. $\eta_{weights}$ represents learning speed. Note in passing that equations (7.4) and (7.5) involve information local to the layers that are being connected. The departure from back-propagation is the presence of an **error signal for the hidden nodes** – this is the crux of the Virtual Targets scheme.

Weights are initialized to random values. The learning scheme then operates as follows:

1. Apply input pattern $\{V_{ip}\}$, and read out the states $\{V_{jp}\}$, $\{V_{kp}\}$ of the hidden and output nodes.
2. Assign targets for the hidden nodes, $\tilde{V}_{jp} = V_{jp}$.
3. Repeat (1), (2) for all input patterns.
4. Present patterns in random order, allowing weights to evolve according to (7.4) and (7.5), and targets $\{\tilde{V}_{jp}\}$ according to:

$$\frac{\delta \tilde{V}_{jp}}{\delta t} = \eta_{targets} \sum_{k=0}^{k=K} \varepsilon_{kp} T_{kj} \tag{7.6}$$

where $\eta_{targets}$ is the target learning speed. In simulation, (7.6) must be multiplied by an additional term of the form $\tilde{V}_{jp}(1-\tilde{V}_{jp})$, to restrain hidden node targets to the range $0 \leqslant \tilde{V}_{jp} \leqslant 1$. When the target values are stored on-chip as charge on capacitors, this 'saturation' will occur naturally. Operations (1)–(3) are 'initialization' steps, and only (4), with equations (7.4)–(7.6), describes learning. During learning via (7.4)–(7.6), whenever the output targets $\{\tilde{V}_{kp}\}$ are achieved for a particular input pattern p (i.e. the net has learned that pattern successfully) the hidden node targets $\{\tilde{V}_{jp}\}$ for that pattern are re-initialized as in (2) above. When a pattern p has been learnt successfully, the errors $\{\varepsilon_{kp}\}$ are, by definition, small. Equations (7.4) and (7.5) will no longer cause weight- and target-modifications, respectively. The target values $\{\tilde{V}_{jp}\}$ may, however, not be equal to the states $\{V_{jp}\}$, and may merely be exerting a force on $\{T_{ji}\}$ via (7.5) in the appropriate direction. It is therefore necessary to introduce this 'reset' mechanism, to cause the learning process to cease to react to pattern p – at least until the weights change to corrupt the successful coding of pattern p.

In the Virtual Targets scheme, the target states $\{\tilde{V}_{jp}\}$ are taking the place of the chain rule in the full back-propagation algorithm, to act as intermediaries in transmitting error information through the intermediate layers. Equations (7.4)–(7.6) describe the learning scheme's response to a single pattern $\{V_{ip}\}$ applied to the inputs, with output states compared to their target values $\{\tilde{V}_{kp}\}$. This is not how MLP training normally proceeds. To mimic conventional MLP learning, each of (7.4)–(7.6) should be allowed to evolve for a short time with each of the input patterns applied, in random order and repeatedly. The simulation experiments in the next section were performed in exactly this way.

7.3.2 Experimental results

The method was applied initially to the standard MLP test problems of Parity- and Encoder-Decoder tasks to determine its functional viability. No fundamental problems were encountered, apart from a tendency to get stuck in local minima, in exactly the same way as in back-propagation learning. To attempt to avoid these minima, noise was injected at the synapse- and activity-levels, with no serious aspiration that learning would survive such rigours. In other words, each synaptic weight $\{T_{ab}\}$ in forward pass mode was augmented by a noise source of variable strength, and each value of $x_a = \Sigma T_{ab} V_b$ was similarly corrupted. Noise sources of up to 20% on either and both of these quantities were introduced. Somewhat surprisingly, learning on these conceptually simple, computationally 'hard' problems was actually **improved** by the presence of high levels of noise, and the network became stuck in local minima less frequently. Including noise in the **backward** calculations (7.4)–(7.6) neither improves nor degrades this result.

Figure 7.1 shows an example of a 4-bit parity learning cycle, with a particularly 'bad' set of (randomly chosen) initial weights. The noise-free network immediately settles into a local minimum, where it stays indefinitely. With noise, however, excursions from the local minimum are made around 2000 and 5300 learning epochs, and a solution is finally found at around 6600 epochs. The temporary minima in the error signal are not associated with one pathologically noisy learning epoch, and the hill-climbing seen in Figure 7.1 takes place over several noisy learning epochs. Learning is clearly enhanced by the presence of noise at a high level (around 20%) on both synaptic weights and activities. This result is surprising, in the light of the normal assertion that back-propagation requires in the order of 16-bit precision during learning, and that analogue VLSI is unsuitable for back-propagation. The distinction is that digital inaccuracy, determined by the significance of the Least Significant Bit (LSB), implies that the smallest possible weight change during learning is 1 LSB. Analogue inaccuracy is,

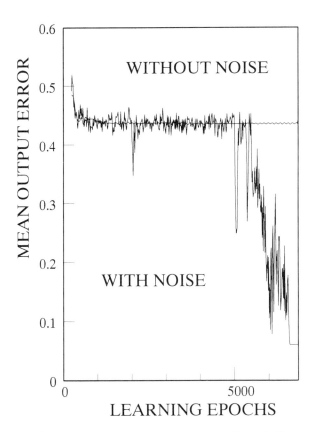

Figure 7.1. *A 4-bit parity learning cycle with a particularly 'bad' set of (randomly chosen) start weights.*

however, fundamentally different, in being noise-limited. In principle, infinitesimally small weight changes **can** be made, and the inaccuracy takes the form of a spread of 'actual' values of that weight as noise enters the forward pass. The underlying 'accurate' weight does, however, maintain its accuracy as a time-average, and the learning process is sufficiently slow to 'see through' the relatively low levels of noise in an analogue system.

The implication is that while analogue noise may introduce inappropriate changes in the $\{T_{ab}\}$ and $\{\tilde{V}_{jp}\}$, the underlying trend reflects the accurate synaptic weight values, and makes the appropriate **averaged** adjustments. The incidental finding – that higher levels of noise actually assist learning – is not so easily explained, although injection of noise into adaptive filter training algorithms is not unusual. These two findings concerning noise would seem to be perfectly general, and have ramifications for all learning processes where weights evolve incrementally, and slowly. In Figure 7.2 the network settles into a poor local minimum shortly after learning commences, with a mean output error of around 0.45. At around 1300 learning epochs, a better, but still local minimum is found (output error $\simeq 0.27$). As the inset shows, the network climbs smoothly out of this local minimum. This effect can be explained as follows. As the network enters the local minimum, a subset of the output patterns are becoming coded correctly (definition of a local minimum). The patterns in this local subset are driving weight changes via (7.4)–(7.6). During this process, the output errors $\{\varepsilon_{kp}\}$

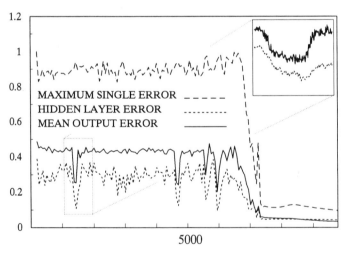

Figure 7.2. *4-bit parity – a local minimum.*

are reducing, as are the hidden node errors $\{\varepsilon_{jp}\}$. Once all of the patterns in the subset are 'learnt', the $\{\varepsilon_{kp}\}$ are all zero, and the target reset mechanism sets the $\{\varepsilon_{jp}\}$ to zero, abruptly. In effect, the hidden node targets, and their associated errors, have been acting as elastic forces on learning, which are suddenly removed. If the local minimum is poor, the input patterns not in the local subset assert themselves via (7.4)–(7.6), and the system climbs out of the poor minimum. Only when the minimum is 'good enough' as defined by the error criterion does it persist – as it does at $\simeq 7000$ epochs in Figure 7.2. This unusual and surprising feature allows the system to respond appropriately to both 'poor' and 'good' minima, as defined by the output error criterion. This behaviour is believed to be a consequence of the target method, coupled with the reset mechanism outlined above.

Parity and Encoder-Decoder tasks are not representative of the real classification/generalization problems that an MLP might aspire to solve. As an example of a more realistic classification task, with both training and test data sets, the Oxford/Alvey vowel database (already described in Section 5.8.5) formed the vehicle for a 54:27:11 MLP to learn to classify vowel sounds from 18 female and 15 male speakers, using the Virtual Targets strategy. The data appear as the analogue outputs of 54 band-pass filters, for 11 different vowel sounds, and 33 speakers. Figure 7.3 shows an example of a learning cycle involving the first five female speakers. The figure charts the evolution of the modulus of the average output error, the hidden node error and the maximum single-bit output error. This latter

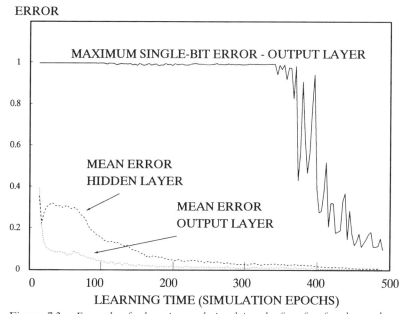

Figure 7.3. *Example of a learning cycle involving the first five female speakers.*

error is included to avoid the situation where the **average** output error is extremely low in the presence of a number of large bit-errors in the output layer, simply because the **number** of outputs is large. Such errors are not acceptable in a classification problem such as this. The maximum single-bit output error is the largest of all the output bit errors, for all nodes and all patterns. Not until it falls can the network be deemed to have learnt the training set completely. Over some 500 simulation (learning) epochs, the errors can be seen to be reduced – not all monotonically. The presence of the noise alluded to above is fairly obvious – and is more so in Figure 7.4, which shows the results of two learning experiments, with and without noise. The experiments were in every other respect identical, using the same set of randomised initial weights. The noise-free traces are smoother, but learning is protracted over the noise-injected example. A solution is found in the absence of noise, and indeed local minima were found to be rare in this set of experiments, with or without noise. The results in Figure 7.4 are, however, dramatic, and characteristic of other similar tests. In each case, learning was ended when the maximum single bit error dropped below 0.1, and the noise signal was reduced in magnitude when this error dropped below 0.4. Interestingly, the generalization ability is also improved by the presence of noise on the synapses and activities – by up to 5%, and the results given above are for a 'noisy' network. The hidden layer errors peak before falling, while the output errors fall more-or-less monotonically. This is entirely consistent with the competing pressures on the hidden node errors.

In an attempt to clarify the role of noise, different levels of noise were applied in a learning cycle with the same initial conditions. The results are shown in Figure 7.5. Initially, learning time is reduced by noise injection, as Figure 7.4 suggests. Increasing the noise level must, however, eventually swamp the data totally, and prevent the classification from being captured at all. This effect is seen in the upper trace in Figure 7.5, where learning times increase exponentially, at a noise level of around 40%. However, the generalization ability (the measure of the quality of learning, as evidenced by the MLP's ability to classify unseen data correctly) rises essentially monotonically. Figure 7.5 suggests that a level of around 10–20% noise offers an optimal compromise between extended learning time for high levels of noise, and lower generalisation ability for lower levels. The most useful observation to be made at this stage is that corrupting the **training data** with noise is held to have the same effect as penalizing high curvature on decision boundaries – in other words, it causes the network to draw sweeping curves through the decision space, rather than fitting convoluted curves to the individual training data points [25]. In this way, underlying trends are modelled, while fine 'irrelevant' detail is ignored.

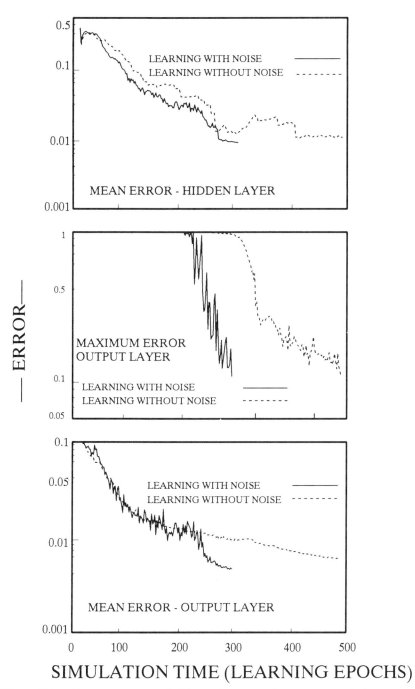

Figure 7.4. *Mean output error, hidden node error and maximum single bit error in two learning experiments, with and without noise.*

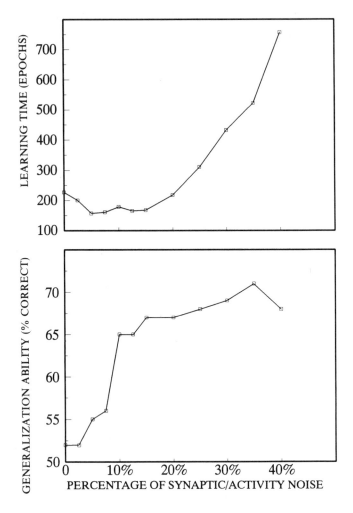

Figure 7.5. *Learning times and generalization ability* versus *noise level injected during learning.*

Noise sources were inserted initially to model the noise known to be present in analogue systems. As the method now stands, the noise sources are being used to improve both learning speed and quality. Analogue noise takes the form of both DC and AC inaccuracies in voltages, currents and device characteristics. DC offsets are essentially cancelled out by including the chip in the learning process, i.e. 'chip-in-the-loop', as described by Intel in the context of their ETANN chip [33] and used with EPSILON in the results described in Section 5.8.5. Natural AC offset noise is in general too weak to provide the high levels of noise that give optimal results here. However, Alspector has reported an elegant solution to this problem in the context of stochastic learning systems [85] involving a small overhead in digital feedback register circuitry. Preliminary experiments suggest that in the work reported in this chapter, the exact form of the noise is not critical, and that a simplified version of Alspector's circuitry will suffice.

The issue of noise is related, clearly, to that of precision and accuracy – always a talking-point in discussion involving analogue VLSI. Its importance, and the surprising, counter-intuitive results shown above justify a more detailed discussion, which is a suitable coda to this book, and is included at the end of this chapter. Once again, we do not include this as an argument in favour of analogue neural VLSI *per se*, but as yet another example where the computational requirements and characteristics of the neural paradigm change the rules for VLSI in a manner that challenges accepted dogma in the digital/analogue debate.

7.3.3 Implementation

Figure 7.6 shows the flow of information in a Virtual Targets network. The forward flow of information is that of a standard MLP. In parallel with this, however, error signals are calculated and passed backwards via equations (7.4)–(7.6). This implies that each synapse circuit must perform backward multiplication as implied by equation (7.6), at the same time as the multiplication for $\Sigma T_{ab} V_b$. We have already shown in Chapters 4 and 5 that multiplication can be performed using as few as three MOSFETs, with the bulk of the synapse circuit devoted to weight storage. The area and

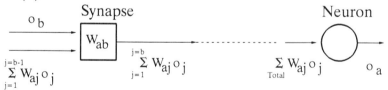

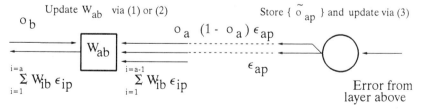

Figure 7.6. *Flow of information in a Virtual Targets network.*

complexity overhead of a 'two-way', single-multiplicand multiplier is therefore slight. Also indicated in Figure 7.6 is the storage requirement for the (adaptive) target states on the hidden units. This could be achieved via a set of on-chip memory elements for each of the $\{\tilde{V}_{jp}\}$. It is more likely to be achieved optimally via a single set of on-chip memories for $\{\tilde{V}_j\}$, loading a new set for each pattern p along with the inputs, and reading the adapted targets $\{\tilde{V}_{jp}\}$ along with the outputs.

The synapse must be capable of incrementing and decrementing its own weight, while passing state signals forward, and error signals backwards. Incrementing and decrementing capacitively stored weights is not difficult [86] and will only involve two or three MOSFETs. The update equation (7.6) for target states is extremely simple, and thus easily implemented local to the neuron circuit.

Looked at critically, this is actually no less **complex** than back-propagation. In fact, the requirement that the current version of the target state be retrieved after each presentation and stored along with the rest of the exemplar from the training set requires an input/output pad for each of the hidden neurons. However, the distinction between hidden and output neurons has been removed to the extent that the update scheme for weights **to** hidden and output neurons is identical, and furthermore both sets of neurons now include a local temporary target storage memory. Chips based on the target method will therefore be architecturally flexible, as the exact role of the neurons in the network's function (input, output or hidden) need not be completely determined at the chip design stage. It is this feature, rather than a significant difference in raw complexity, that renders this scheme more amenable to VLSI.

7.4 The bottom-up approach: weight perturbation

The most direct form of hardware learning is a scheme whereby each of the weights is perturbed *one at a time* and the effect on the output error is measured in order to update that particular weight. Weight perturbation [87] is an exact mathematical equivalent to error back-propagation, but it is much more amenable to hardware implementation. No assumptions about the details of the circuitry are required: for example, knowledge of the transfer function of each neuron is not needed. Weight perturbation is a finite-difference approximation scheme for evaluating $\partial E/\partial T_{kj}$. A perturbation $\Delta\tau$ is applied to a weight T_{kj} and the change in mean square error as a result of this perturbation ΔE is measured. The value of $\partial E/\partial T_{kj}$ is then estimated as follows:

$$\frac{\partial E}{\partial T_{kj}} \approx \frac{\Delta E}{\Delta \tau} \quad \text{and} \quad \text{hence} \quad \Delta T_{kj} = -\eta \frac{\Delta E}{\Delta \tau} \tag{7.7}$$

In the limit, as $\Delta\tau \to 0$, back-propagation and weight perturbation are exactly equivalent. With weight perturbation, however, the weight update is simply a scalar multiple of the change in E, since both η and $\Delta\tau$ are constants.

The main motivation for wanting to use weight perturbation is its extreme simplicity. As already stated, our aim is to design a learning scheme which does *not* require the use of a host processor. Weight perturbation as a learning procedure consists of the repeated application of four very simple steps: a weight T_{kj} has an amount $\Delta\tau$ temporarily added to it and the change in output error ΔE is measured. The appropriate scaling factor is applied to obtain ΔT_{kj} (equation (7.7)) which is then stored in memory before $\Delta\tau$ is removed. The perturbation is then applied to the next weight until eventually all weight updates have been computed for the same input pattern. At this point, all the weights are updated at once before the next pattern is presented.

To assess the method of weight perturbation for learning with analogue hardware, we carried out a software feasibility study taking into account the limitations of analogue circuitry as rigorously as possible. A reasonably complex test problem was constructed for this study as results obtained on toy problems can often be misleading. Simulations of precision, noise or leakage were also based on measurements made on the VLSI circuits previously described in Chapter 5.

7.5 Test problem

We constructed a synthetic classification problem using our work on mobile robot navigation, so that we could have full control over the process of data generation. An idealized 2D environment was created, consisting of an L-shaped room with six corners and two rectangular obstacles (Figure 7.7).

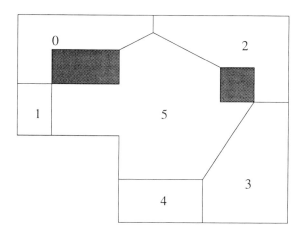

Figure 7.7. *Idealized 2D environment for robot localization.*

A simulator was written to generate 360° range scans from any (x, y) coordinates within this environment. The set of range values which make up the scan was treated purely as a pattern from which the following eight features were extracted: the shortest, median and longest ray lengths, the energy in the pattern (i.e. sum of the squared ray lengths), the size of the two largest discontinuities in the pattern, the angle subtended by the longest wall segment and the perpendicular distance from it (the last two features requiring a reliable corner detection algorithm). To turn the pattern processing task into one of classification, the room was divided into six regions using the nearest visible corner as the criterion for tessellation (see Figure 7.7). A database consisting of several thousand input–output pairs generated from random locations throughout the room was assembled.

To obtain a suitable benchmark with which to compare subsequent simulations of *hardware* learning, an 8-25-6 multi-layer perceptron was trained with back-propagation using full 64-bit precision. The plot of test error rate *versus* number of hidden units has a very shallow minimum for this problem, and so the choice of a value of 25 for the number of hidden units is not critical. The input data is perfect in the sense that it is noiseless and that all the training patterns are correctly labelled. When the training set consists of as many as 5000 patterns, we can expect the multi-dimensional curve-fitting achieved by the network to be very accurate. The classification of the test set, which consists of a further 500 patterns randomly chosen from the database, should be an interpolation task for nearly all of these patterns and hence the test error rate will be very low under these conditions, as is shown by the results of Table 7.1 (error rates less than 3% for 5000 training patterns).

A direct comparison of the performance of weight perturbation with full 64-bit precision and a value of 0.0001 for $\Delta \tau$ was next carried out on the same problem, using the 500-pattern training set to keep the simulations runs to reasonable lengths. The results are shown in Table 7.2, which also gives the values of the mean and standard deviation for the weight sets. In

Table 7.1 Classification error rate on region identification problem for 8-25-6 networks trained with back-propagation. An iteration is a single presentation (in random order) of all the patterns in the training set. All the runs were carried out with a learning rate of 0.2 but with no momentum term

Number of training patterns	Number of Iterations		
	500	1000	2000
500	11.58%	8.07%	7.10%
1000	7.14%	5.71%	5.31%
5000	2.86%	2.50%	2.42%

Table 7.2 Comparison of back-propagation and weight perturbation on region identification problem for 8-25-6 networks. In all cases the learning rate η is set at 0.2.

Method	No. of iterations	Error rate	Mean weight	s.d. for weight set
(a) Back-propagation	500	11.58%	−0.4473	3.9008
(b) Back-propagation	1000	8.42%	−0.5773	5.4590
(c) Weight perturbation	500	11.58%	−0.4471	3.9012
(d) Weight perturbation	1000	8.42%	−0.5771	5.4604

all cases, the value of η was again set at 0.2, and the same initial random weights were used in all experiments. The error rates in Table 7.2 (as with all the remaining results in this chapter) represent the error obtained on the 500-pattern test set averaged over the last 100 iterations of the learning procedure. This enables us to compare the performance of each learning procedure under exactly the same conditions. As expected, the results of Table 7.2 show that error back-propagation and weight perturbation are exactly equivalent when $\Delta\tau$ is sufficiently small.

7.6 Weight perturbation for hardware learning

We now come to the key part of this section. If weight perturbation (or any other learning strategy, for that matter) is to be used for on-chip learning with analogue VLSI, it can only work *within the limitations of analogue hardware*. We saw in Chapter 5 that changes in weight values smaller than 1 part in 1000 are lost in the noise and this introduces the most critical constraint on any learning scheme for analogue VLSI: weights cannot effectively be set to a precision better than 1 part in 1000. This, together with the requirement on the dynamic range of the weight set, determines the size of the smallest weight update possible.

The first choice to be made in our simulations of learning within the limitations of analogue hardware is that of the dynamic range of the synaptic weights. For reasons that will soon become apparent, we chose to limit the range of weight values to be between −10 and +10 for this problem. Under these conditions, the smallest weight update possible is 0.02 (0.1% of 20), i.e. a weight can only be incremented or decremented in multiples of 0.02. The next parameter to be modified is the value of $\Delta\tau$; it is sensible to choose a perturbation which is greater than the smallest value possible (0.02) and we therefore used a value of 0.1 in our simulations. Each hardware constraint (increase in size of perturbation to 0.1, weight clipping and weight update quantization[2] is introduced in turn in the simulations. An increase in the size of the applied perturbation from 0.0001 to 0.1 actually gives a slight improvement in performance (see A and B in Table 7.3). With such a large perturbation, weight perturbation is no longer a true gradient descent procedure; improved performance has been observed by others – see, for example [84], with non-gradient descent techniques. Here it may be due to the use of a greater range of synaptic weights giving better discrimination on a hard classification problem. The classification test error rate decreases to 10.36% whilst the standard deviation of the weight set increases to 5.0082. With a standard deviation of this value, 95% of the weights will lie between ±10.0 of the mean value, and this was the reason for the choice of a weight range extending from −10 to +10. When weight clipping according to this criterion is introduced in the next simulation (C in Table 7.3), the test error rate only rises by 1.3%.

Table 7.3 The effect of introducing realistic hardware constraints on the performance of weight perturbation. Note that the input data and the output values from the hidden units are also encoded with 0.1% precision, but the effect of this is negligible with respect to limited **weight** precision

	Size of perturbation	Weight range	Weight update in steps of 0.02	Learning rate	Error rate (%)	s.d. for weight set
A	0.0001	No limit	No	0.2	11.58	3.9050
B	0.1	No limit	No	0.2	10.36	5.0082
C	0.1	−10 to +10	No	0.2	11.68	4.0617
D	0.1	−10 to +10	Yes	0.2	54.86	1.2850
E	0.1	−10 to +10	Yes	1.0	14.60	5.2798

Not surprisingly, however, the main effect occurs when the weight update 'quantization' is introduced (weight updates in integer multiples of 0.02 whilst keeping the 0.1 perturbation and the ±10 limits). As a result of including this further modification, we found that learning was no longer possible (test error rate greater than 50%). Closer examination of the data revealed that 99% of the $|\Delta T_{kj}|$ values were less than 0.02 and were therefore being set to zero. When the value of η was increased to 1.0, however, a test classification error rate of 14.60% was obtained, i.e. a degradation of only 2.9% from C to E in Table 7.3. As many as 92% of the ΔT_{kj} values are still being set to zero; nevertheless, there are enough weights being updated for learning to proceed.

[2] Strictly speaking, all weight updates are quantized with digital simulations but with 64-bit double precision, the size of the quantization step is negligible.

Results D and E in Table 7.3 prompted us to undertake a more thorough investigation of the effect on the error rate of varying η. The results, shown in Figure 7.8, indicate not only that a value of 1.0 for η is the best in this case, but also that any value between 0.6 and 2.4 would be acceptable for this test problem.

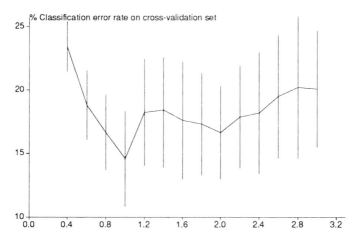

Figure 7.8. *Error rate on cross-validation set as a function of the learning rate.*

The results given in this section demonstrate that on-chip learning with weight perturbation is potentially feasible with analogue VLSI if the learning rate is carefully chosen. We do need to introduce a note of caution at this stage, however: the error rate is very sensitive to the actual values in the weight set. This can be demonstrated by a plot of the change in classification error rate (*not* the error rate E on the training set) over time when testing on a cross-validation set during learning. Figure 7.9 shows such a plot for cases A and E in Table 7.3, starting from the same set of random weights. Note the lack of smoothness of the curve for E, showing large variations from one iteration to the next. This suggests the use of an altogether different criterion to decide when to terminate the learning phase and the need to store, during learning, the weight set corresponding to the minimum cross-validation error so far achieved.

7.7 Back-propagation revisited (for the final time?)

We are now in a position to comment on whether back-propagation is indeed possible with analogue hardware, a subject of some controversy as indicated at the beginning of this chapter. To do this, we investigate the error back-propagation algorithm *under exactly the same conditions* as we have just tested weight perturbation: weight updates in multiples of 0.02 for

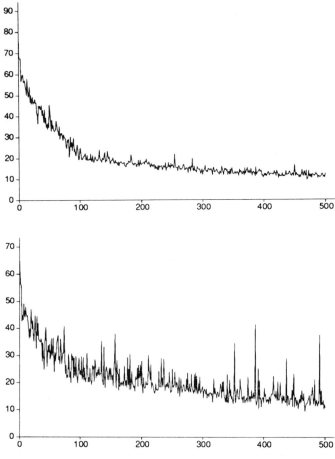

Figure 7.9. *Error rate on cross-validation set for cases A and E.*

an overall dynamic range of ± 10. With error back-propagation, the problem is complicated by the fact that there are three or four successive products to be evaluated to compute the weight updates (see equations (7.2) and (7.3)). If learning is to take place on chip, an analogue multiplier is required for each of these products and the precision of each multiplier circuit has to be taken into account. As a result, we are forced to make a number of further assumptions in our simulations of error back-propagation as a hardware learning scheme. We assume that each of the input variables is known to a precision of 1 part in 1000 and further that the result of each successive multiplication is also available with the same precision. This latter assumption is very much a best-case assumption: if two variables x and y are encoded with an error ε, their product will have an error term of the form $(x+y)\varepsilon$. If $-1 \leqslant x, y \leqslant 1$, as is the case for most of the terms in equations (7.3) and (7.4), the error could be as much as 2ε although, on average, it would be much less. This reasoning, however, does *not* include the inaccuracy introduced by the analogue multiplier itself.

The variables D_k and V_k in equation (7.2) lie in the range 0 to 1 which means that δ_k can take on any value between ± 1. T_{kj} in equation (7.3) can vary between ± 10, but the value of the $\Sigma \delta_k T_{kj}$ term cannot be predicted and hence the dynamic range of δ_j or ΔT_{ji} (for the first layer) is not known *a priori*. In our simulations, ΔT_{ab} was therefore allowed to take on any value between -1 and $+1$, in steps of 0.02, as before. When back-propagation was simulated under these conditions, the results of Figure 7.10 were obtained. These results are equivalent to those of Figure 7.8: they represent the classification error rate on the same cross-validation set as the learning rate η is varied from 0.2–2.4. The results show that there is also an optimal learning rate ($\eta = 1.2$) for which the error rate is 17.8%, i.e. a further 3% degradation in performance when compared with weight perturbation.

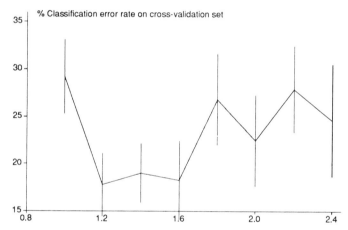

Figure 7.10. *Error rate on cross-validation set as a function of the learning rate – back-propagation.*

It would appear that, on this hard benchmark problem, back-propagation learning does remain possible once the learning rate is optimized, although the increase in classification error rate (6%) due to the limitations of analogue hardware was twice that obtained with weight perturbation. This result would seem to support the results of Dolenko and Card [83] who claim that back-propagation learning is possible with analogue circuitry. However, it is our opinion that, whilst back-propagation may *theoretically* be feasible with analogue VLSI, it is unlikely to work in practice. We base this remark on three facts: firstly, we have assumed in our simulations that the analogue circuitry is working at the limits of its performance (individual weights settable to a precision of 1 part in 1000). More importantly perhaps, the cascading of four analogue multipliers is unlikely to produce an answer which is also accurate to 1 part in 1000. Finally, the routing of all the input and intermediate data required for the computation of that answer is a formidable obstacle to a practical realisation. Since weight perturbation is unaffected by the last two constraints, we have no doubt that it is the type of learning scheme which should be adopted for analogue VLSI.

7.8 Conclusion

Weight perturbation has been shown in this chapter to be an appropriate learning scheme for on-chip implementation in analogue VLSI. The degradation in its performance caused by the introduction of realistic hardware constraints is acceptable and the extra hardware requirements for on-chip learning are minimal. Since η, the learning rate, and $\Delta\tau$, the size of the perturbation, are constant, the weight change ΔT_{ab} is simply a (constant) scalar multiple of the change in error, ΔE. Thus the hardware needed to implement on-chip weight perturbation is as follows:

- a means of adding an offset to each weight voltage
- a module to measure the chip outputs, compare them with desired values[3] to evaluate the change in output error ΔE and scale it for the computation of ΔT_{ab}
- a means of storing the weight update and subsequently adding it to the current weight value.

The results presented in this section assume that individual weight values can be set to a precision of 1 part in 1000 and therefore represent the *best* performance which could be obtained with analogue technology. If we have to relax this criterion by assuming instead that the weights can only be set to 1 part in 250 (i.e. equivalent to 8-bit precision), our simulations indicate that this has a catastrophic effect and learning is no longer possible with weight perturbation. There is a solution available to this problem, however: *probabilistic* weight updates, as recently suggested [88]. If a weight update $|\Delta T_{ab}|$ is less than the smaller weight update possible $|\Delta T_{ab}|_{min}$ (0.08 for the case of 8-bit precision on our test problem), the probability that an update of ± 0.08 takes place depends on the ratio $|\Delta T_{ab}|/|\Delta T_{ab}|_{min}$. Thus, if $|\Delta T_{ab}|$ was equal to, say, 0.01, the weight T_{ab} would be updated in 1 out of 8, or 12.5%, of cases. When we implemented this scheme[4] on the test problem described in this section, we found that learning once again became possible, and Table 4 shows that 8-bit equi-

---

[3] These would probably be stored in off-chip memory.
[4] Which, admittedly, carries a further hardware overhead although maximal length pseudo-random binary sequences can easily be generated in hardware using a shift register and an exclusive-OR gate.

Table 7.4 The extra entry (F) shows the effect of updating weights probabilistically with 8-bit precision only

	Size of perturbation	Weight range	Weight update in steps of 0.02	Learning rate	Error rate (%)	s.d. for weight set
A	0.001	No limit	No	0.2	11.58	3.9050
B	0.1	No limit	No	0.2	10.36	5.0082
C	0.1	-10 to $+10$	No	0.2	11.68	4.0617
D	0.1	-10 to $+10$	Yes	0.2	54.86	1.2850
E	0.1	-10 to $+10$	Yes	1.0	14.60	5.2798
F	0.1	-10 to $+10$	0.08	1.0	17.74	5.9833

valent *probabilistic* updates (F) give results almost identical to the 10-bit equivalent updates (E).

The main disadvantage with weight perturbation is the slowness with which learning proceeds. With error back-propagation, a single output error measurement is required to compute the δ's for each layer and adjust all of the weights. With the method of weight perturbation, one error measurement is required *every time a weight is perturbed*; the learning time is therefore proportional to the number of weights in the network. An intermediate solution is the technique of node perturbation – sometimes known as Madaline Rule III [9], whereby it is the activity x of each neuron which is perturbed and δ is estimated from measurement of $\Delta E/\Delta x$. In this case, the learning time is proportional to the number of neurons in the network but extra hardware is now required to apply the Δx perturbation and to compute the $\delta_k V_j$ products. The method does therefore have a significant hardware overhead when compared with weight perturbation and it may be better instead to try and develop *parallel* weight perturbation methods in which all the weights to a node are perturbed simultaneously [89]. It is our opinion also that such variants of weight perturbation will make it possible to achieve on-chip learning with analogue VLSI in reasonable time.

7.9 Noisy synaptic arithmetic – an analysis

Co-Author: Peter Edwards

In this concluding section, we demonstrate both by consideration of the cost function and the learning equations, and by simulation experiments, that injection of random noise on to MLP weights during learning enhances fault-tolerance without additional supervision. We also show that the nature of the hidden node states and the learning trajectory is altered fundamentally, in a manner that improves training times and learning quality. The enhancement uses the mediating influence of noise to distribute information optimally across the existing weights.

Taylor [90] has studied noisy synapses, largely in a biological context, and infers that the noise might assist learning. We have already demonstrated that noise injection both reduces the learning time and improves the network's generalization ability. Here we infer (synaptic) noise-mediated terms that sculpt the error function to favour faster learning, and that generate more robust internal representations, giving rise to better generalization and immunity to small variations in the characteristics of the test data. Much closer to the spirit of this section is the work of Hanson [91]. His stochastic version of the delta rule effectively adapts weight means and standard deviations. Also Sequin and Clay [92] use stuck-at faults during training which imbues the trained network with an ability to withstand such faults. They also note, but do not pursue, an increased generalization ability.

This section presents an outline of the mathematical predictions and verification simulations. A full description of the work is given in [93].

7.9.1 Mathematical predictions

Let us analyse an MLP with I input, J hidden and K output nodes, with a set of P training input vectors $Q^p = \{V_{ip}\}$, looking at the effect of noise injection into the error function itself. We are thus able to infer, from the additional terms introduced by noise, the characteristics of solutions that tend to **reduce** the error, and those which tend to **increase** it. The former will clearly be favoured, or at least stabilised, by the additional terms, while the latter will be de-stabilised.

Let each weight T_{ab} be augmented by a random noise source, such that $T_{ab} \to T_{ab} + \Delta_{ab} T_{ab}$, for all weights $\{T_{ab}\}$. Neuron thresholds are treated in precisely the same way. Note in passing, but importantly, that this **synaptic** noise is **not** the same as noise on the input data. Input noise is correlated across the synapses leaving an input node, while the synaptic noise that forms the basis of this study is not. The effect is thus quite distinct.

Considering, therefore, an error function of the form:

$$\varepsilon_{tot,p} = \frac{1}{2} \sum_{k=0}^{K-1} \varepsilon_{kp}^2 = \frac{1}{2} \sum_{k=0}^{K-1} (V_{kp}(\{T_{ab}\}) - \tilde{o}_{kp})^2 \tag{7.8}$$

where $\tilde{o}_{kp}$ is the target output. We can now perform a Taylor expansion of the output V_{kp} to second order, around the noise-free weight set $\{T_N\}$, and thus augment the error function:

$$V_{kp} \to V_{kp} + \sum_{ab} T_{ab} \Delta_{ab} \left(\frac{\partial V_{kp}}{\partial T_{ab}} \right)$$

$$+ \frac{1}{2} \sum_{ab,cd} T_{ab} \Delta_{ab} T_{cd} \Delta_{cd} \left(\frac{\partial^2 V_{kp}}{\partial T_{ab} \partial T_{cd}} \right) + O(>3) \tag{7.9}$$

If we ignore terms of order Δ^3 and above, and taking the time average over the learning phase, we can infer that two terms are added to the error function:

$$\langle \varepsilon_{tot} \rangle = \langle \varepsilon_{tot}(\{T_N\}) \rangle$$

$$+ \frac{1}{2P} \sum_{p=1}^{P} \sum_{k=0}^{K-1} \Delta^2 \sum_{ab} T_{ab}^2 \left[\left(\frac{\partial V_{kp}}{\partial T_{ab}} \right)^2 + \varepsilon_{kp} \left(\frac{\partial^2 V_{kp}}{\partial T_{ab}^2} \right) \right] \quad (7.10)$$

Also the perceptron rule update on the hidden-output layer along with the expanded error function becomes:

$$\langle \delta T_{kj} \rangle = -\tau \sum_p \langle \varepsilon_{kp} V_{jp} o'_{kp} \rangle - \tau \frac{\Delta^2}{2} \sum_P \langle V_{jp} o'_{kp} \rangle \times \sum_{ab} T_{ab}^2 \frac{\partial^2 V_{kp}}{\partial T_{ab}^2} \quad (7.11)$$

averaged over several training epochs (which is acceptable for small values of τ the *adaption rate* parameter).

7.9.2 Simulations

Two contrasting classification tasks were selected to verify the predictions made in the following section by simulation. The first, a feature location task, uses *real world* normalized grey scale image data. The task was to locate eyes in facial images – to classify sections of these as either 'eye' or 'not-eye'. The network was trained on 16×16 preclassified sections of the images, classified as eyes and not-eyes. The not-eyes were random sections of facial images, avoiding the eyes (see Figure 7.11). The second, a more artificial task, was the ubiquitous character encoder (Figure 7.12) where a 25-dimensional binary input vector describing the 26 alphabetic characters (each 5×5 pixels) was used to train the network with a one-out-of-26 output code.

During the simulations noise was added to the weights at a level proportional to the weight size and at a probability distribution of uniform density (i.e. $-\Delta_{max} < \Delta < \Delta_{max}$). Levels of up to 40% were probed in detail – although it is clear that the expansion above is not quantitatively valid at

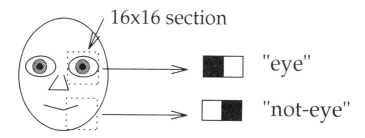

Figure 7.11. *Eye/not-eye classification task – schematic.*

Input Patterns

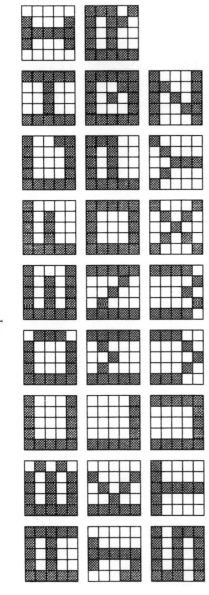

Output Targets

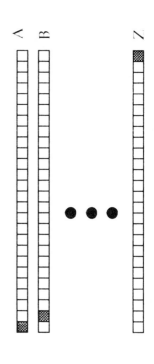

Figure 7.12. *Character encoder classification task – schematic.*

this level. Above these percentages further improvements were seen in the network performance, although the dynamics of the training algorithm became chaotic. The injected noise level was reduced smoothly to a minimum value of 1% as the network approached convergence (as evidenced by the highest output bit error). As in all neural network simulations, the results depended upon the training parameters, network sizes and the random start position of the network. To overcome these factors and to achieve a meaningful result 35 weight sets were produced for each noise level. All other characteristics of the training process were held constant. The results are therefore not simply pathological cases.

7.9.3 Prediction/verification

Fault tolerance
The first derivative penalty term in the expanded cost function (7.10) averaged over all patterns, output nodes and weights, becomes:

$$K \times \Delta^2 \left[T_{ab}^2 \left(\frac{\partial V_{kp}}{\partial T_{ab}} \right)^2 \right] \tag{7.12}$$

The implications of this term are straightforward. For large values of the (weighted) average magnitude of the derivative, the overall error is increased. This term therefore causes solutions to be favoured where the dependence of outputs on individual weights is evenly distributed across the entire weight set. Furthermore, weight saliency should not only have a lower average value, but a smaller scatter across the weight set as the training process attempts to reconcile the competing pressures to reduce both (7.8) and (7.12). This more distributed representation should be manifest in an improved tolerance to faulty weights.

Simulations were carried out on 35 weight sets produced for each of the two problems at each of five levels of noise injected during training. Weights were then randomly removed and the networks tested on the training data. The resulting graphs (Figures 7.13, 7.14) show graceful degradation with an increased tolerance to faults with injected noise during training. The networks were highly constrained for these simulations to remove some of the natural redundancy of the MLP structure. Although the eye/not-eye problem contains a high proportion of redundant information, the improvement in the network's ability to withstand damage, with injected noise, is clear.

7.9.4 Generalization ability

Considering the derivative in equation (7.12), and looking at the input-hidden weights, the term that is added to the error function, again averaged

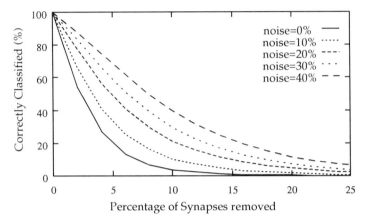

Figure 7.13. *Graceful degradation with synaptic destruction – character encoder task with noisy training.*

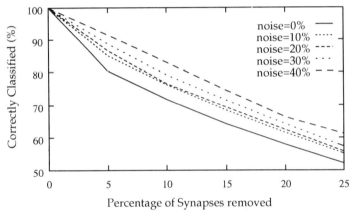

Figure 7.14. *Graceful degradation with synaptic destruction – eye/not-eye classifier with noisy training.*

over all patterns, output nodes and weights is:

$$K \times \Delta^2 [T_{ji}^2 o'_{kp}^2 T_{kj}^2 o'_{jp}^2 V_{ip}^2] \tag{7.13}$$

If an output neuron has a non-zero connection from a particular hidden node ($T_{kj} \neq 0$), and provided the input V_{ip} is non-zero and is connected to the hidden node ($T_{ji} \neq 0$), there is also a term o'_{jp} that will **tend to favour solutions with the hidden nodes also turned firmly ON or OFF** (i.e. $V_{jp} = 0$ *or* 1). Remembering, of course, that all these terms are noise-mediated, and that during the early stages of training, the 'actual' error ε_{kp}, in (7.8) will dominate, this term will de-stabilize final solutions that balance the hidden nodes on the slope of the sigmoid. Naturally, hidden nodes V_j that are firmly

ON or OFF are less likely to change state as a result of small variations in the input data $\{V_i\}$. This should become evident in an increased tolerance to input perturbations and therefore an increased generalization ability.

Simulations were again carried out on the two problems using 35 weight sets for each level of injected synaptic noise during training. For the character encoder problem generalization is not really an issue, but it is possible to verify the above prediction by introducing random Gaussian noise into the input data and noting the degradation in performance. The results of these simulations are shown in Figure (7.15), and show a clearly increased ability to withstand input perturbation, with injected noise into the synapses during training.

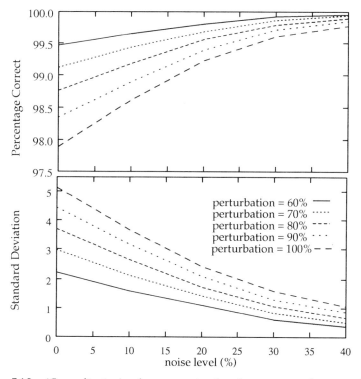

Figure 7.15. *'Generalization' enhancement in the character encoder – artificially generated test data.*

Generalization ability for the eye/not-eye problem is a real issue. This problem therefore gives a valid test of whether the synaptic noise technique actually improves generalization performance. The networks were therefore tested on previously unseen facial images and the results are shown in Table 7.5.

Table 7.5 Generalization enhancement shown through increased ability to classify previously unseen data in the eye/not-eye task

Correctly classified (%)					
Noise levels	0%	10%	20%	30%	40%
Test patterns	67.875	70.406	70.416	72.454	75.446

These results show dramatically improved generalization ability with increased levels of injected synaptic noise during training. An improvement of approximately 8% is seen – consistent with earlier results on the different 'real' problem demonstrated at the beginning of this chapter, and in [72].

7.9.5 Learning trajectory

The second derivative penalty term in the expanded cost function (7.9) is perhaps even more intriguing. This term is complex as it involves second order derivatives, and also depends upon the sign and magnitude of the errors themselves $\{\varepsilon_{kp}\}$. The simplest way of looking at its effect is to look at a single exemplar term:

$$K\Delta^2 \varepsilon_{kp} T_{ab}^2 \left(\frac{\partial^2 V_{kp}}{\partial T_{ab}^2}\right) \qquad (7.14)$$

This term implies that when the combination of $\varepsilon_{kp}\partial^2 V_{kp}/\partial T_{ab}^2$ is negative then the overall cost function error is reduced, and *vice versa*. Term (7.14) is therefore constructive as it can actually lower the error locally via noise injection, whereas (7.13) always increases it. Term (7.14) can therefore be viewed as a sculpting of the error surface during the early phases of training (i.e. when ε_{kp} is substantial). In particular, a weight set with a higher 'raw' error value, calculated from (7.8), may be favoured over one with a lower value if noise-injected terms indicate that the 'poorer' solution is located in a promising area of weight space. This 'look-ahead' property should lead to an enhanced learning trajectory, perhaps finding a solution more rapidly.

In the augmented weight update equation (7.11) the noise is acting as a medium, projecting statistical information about the character of the entire weight set on to the update equation for each particular weight. So, the effect of the noise term is to account not only for the weight currently being updated, but to add in a term that estimates what the other weight changes are likely to do to the output, and adjust the size of the weight increment/decrement as appropriate.

To verify this by simulation is not as straightforward as the other predictions. It is, however, possible to show the mean training time for each level of injected noise. For each noise level, 1000 random start points

NOISE IN TRAINING – SOME CONCLUSIONS

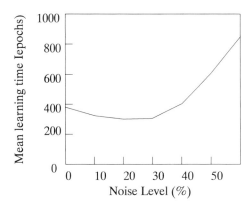

Figure 7.16. *Training time against noise injected into training.*

were used to allow the underlying properties of the training process to emerge. The results are shown in Figure 7.16, and clearly show that at low noise levels ($\leqslant 30\%$ for the case of the character encoder) a definite reduction in training times are seen. At higher levels the chaotic nature of the 'noisy learning' takes over.

It is also possible to plot the combination of $\varepsilon_{kp}(\partial^2 V_{kp}/\partial T_{ab}^2)$. This is shown in Figure 7.17, again for the character encoder problem. Term (7.14) is reduced more quickly with injected noise, thus effecting better weight changes. At levels of noise $>7\%$ the effect is exaggerated, and the noise mediated improvements take place during the first 100–200 epochs of training. The level of 7% is displayed

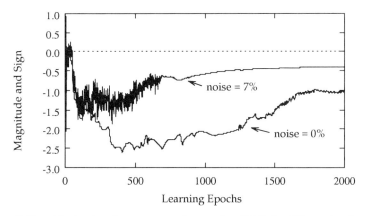

Figure 7.17. *Monitoring the extra term in equation (7.14) for different levels of noise injection.*

simply because it is visually clear what is happening, and is also typical.

7.10 Noise in training – some conclusions

Many potential neural network applications can be implemented as software, digital or analogue hardware, where the learning phase is not part of the final 'product'. If, however, the application makes overt use of the *adaptive* facet of the neural paradigm, the learning algorithm or strategy impacts directly on the details of the implementation (and *vice versa*). For instance, if it is necessary either to configure the network for a set of unique circumstances, or to adapt to changing circumstances, the 'product' must incorporate learning software or hardware.

7.11 On-chip learning – conclusion

Adaptive neural systems with on-chip learning will be useful in any context where the task for which they were initially trained has to be switched, for one reason or another. They will be especially useful in vision applications for which massive parallelism is often required to achieve real-time processing, for example the tracking of a solid object in real time using a video camera. We have presented, in this chapter, two attempts to develop hardware learning strategies for multi-layer perceptrons. They start from diametrically opposite sets of assumptions and neither can yet be said to be the ultimate on-chip learning strategy. Nevertheless, results are promising and some of the commonly-held wisdom about accuracy in analogue computation has been challenged. The work on the Virtual Targets algorithm has revealed some intriguing effects thrown up by the departure, of necessity, from an algorithm which aims for pure gradient descent. As with most of the work described in this book, our findings have reinforced our view that pragmatic considerations and an occasional look at *real* neural networks are as fruitful in the development of artificial neural networks as the application of mathematical rigour.

We have also shown both by mathematical expansion and by simulation that injecting random noise on to the synaptic weights of a multilayer perceptron during the training phase enhances fault-tolerance, generalization ability and learning trajectory. It has long been held that any inaccuracy during training is detrimental to MLP learning. We have proved that **analogue** inaccuracy is not. The mathematical predictions are perfectly general and the simulations relate to a non-trivial classification task and a 'real' world problem. The results are therefore important for the designers of analogue hardware and also as a non-invasive technique for producing learning enhancements in the software domain.

References

1. Hopfield, J. J. (1982) *Neural Networks and Physical Systems with Emergent Collective Computational Abilities.* Proceedings of the National Academy of Science USA, **79**, 2554–8, April.
2. Hopfield, J. J. (1984) *Neural Networks and Physical Systems with Graded Response have Collective Properties like those of Two-State Neurons.* Proceedings of the National Academy of Science USA, **81**, 3088–92, May.
3. Graf, H. P., Jackel, L. D., Howard, R. E., *et al.* (1986) *VLSI Implementation of a Neural Network Memory with Several Hundreds of Neurons.* Proceedings of the AIP Conference on Neural Networks for Computing, Snowbird, UT, 182–7.
4. Tarassenko, L., Murray, A. F. and Tombs, J. (1989) *Neural Network Architectures for Associative Memory.* IEE Conference on Artificial Neural Networks, 17–22.
5. McEliece, R. J., Posner, E. C., Rodemich, E. R. and Venkatesh, S. S. (1987) The capacity of the Hopfield associative memory. *IEEE Trans. Infor. Theory*, **33**, 461–82.
6. Amit, D. J., Gutfreund, H. and Sompolinsky, H. (1985) Storing infinite numbers of patterns in a spin glass model of neural networks. *Phys. Rev. Lett.*, **55**, 1530–3.
7. Tarassenko, L., Tombs, J. and Reynolds, J. H. (1991) Neural network architectures for content-addressable memory. *IEE Proc. Pt. F*, **138**(1), 3–39.
8. Rumelhart, D. E. and McLelland, J. D. (1986) *Parallel Distributed Processing: Explorations in the Microstructures of Cognition Volume 1*, MIT Press, Cambridge, MA.
9. Widrow, B. and Lehr, M. A., (1990) 30 years of adaptive neural networks: Perceptron, Madaline, and backpropagation. *Proc. IEEE*, **78**, 1415–42.
10. Rosenblatt, F. (1958) The Perceptron: A probabilistic model for information storage and organisation in the brain. *Psychol. Rev.*, **65**, 386–408.
11. Minsky, M. L. and Papert, S. A. (1969) *Perceptrons: An Introduction to Computational Geometry*, MIT Press, Cambridge, MA.
12. Darken, C. and Moody, J. (1991) *Note on Learning Rate Schedules for Stochastic Optimisation.* Neural Information Processing Systems (NIPS) Conference, 832–8, Morgan-Kaufmann, San Mateo, CA.
13. Kohonen, T. (1984) *Self-organisation and Associative Memory*, Springer-Verlag, Berlin.

14. Anderson, J., Kirk, D. and Platt, J. (1993) *An Analog VLSI Chip for Radial Basis Functions*. Neural Information Processing Systems (NIPS) Conference, 765–772, Morgan-Kaufmann, San Mateo, CA.
15. Lippmann, R. P. (1987) An introduction to computing with neural nets. *IEEE ASSP Mag.*, April, 4–22.
16. Lowe, D. (1991) *What have Neural Networks to offer Statistical Pattern Processing?* SPIE Conference on Adaptive Signal Processing, San Diego, CA, July.
17. Rao, A., Walker, M. R., Clark, L. T. and Akers, L. (1989) *Integrated Circuit Emulation of ARTI Networks*. 1st IEE Conference on Artificial Neural Networks, 37–41.
18. Mead, C. and Conway, L. (1981) *Introduction to VLSI Systems*. Addison-Wesley, Reading, MA.
19. Allen, P. E. and Holberg, D. R. (1987) *CMOS Analog Circuit Design*, Holt, Rinehart and Winston, New York.
20. Sze, S. (1981) *Physics of Semiconductor Devices*, Wiley, New York, NY.
21. Hammerstrom, D. (1991) A highly parallel digital architecture for neural network emulation, in: *VLSI for AI and Neural Networks*, 357–366, Moor, W. R. and Delgado-Frias, J. D. Plenum Press, New York, NY.
22. Le Cun, Y., Jackel, L. D., Graf, H. P. *et al.* (1990) *Optical Character Recognition and Neural-Net Chips*. Proceedings of the International Neural Network Conference (INNC-90), Paris, France, 651–5.
23. Mead, C. (1988) *Analog VLSI and Neural Systems*, Addison-Wesley, Reading, MA.
24. Vittoz, E. (1985) Micropower techniques, in *Design of MOS VLSI Circuits for Telecommunications* (eds. Y. Tsividis and P. Angonetti), Prentice-Hall, Englewood Cliffs, NJ.
25. Bishop, C. (1990) *Curvature-Driven Smoothing in Backpropagation Neural Networks*. Proceedings of the International Neural Networks Conference (INNC-90), Paris, France, 749–52.
26. Murray, A. F. (1991) Analogue noise-enhanced learning in neural network circuits. *Electr. Lett.*, **2**(17), 1546–8.
27. Holmstrom, L. and Koistinen, P. (1992) Using additive noise in back-propagation training. *IEEE Trans. Neural Networks*, **3**(1), 24–38.
28. Gilbert, B. A. (1968) Precise four-quadrant multiplier with sub-nanosecond response. *IEEE J. Solid-State Circ.* **3**, 365–73.
29. Ryckebusch, S., Mead, C. and Bower, J. (1988) *Modelling small oscillating Biological Networks in Analog VLSI*. Neural Information Processing Systems (NIPS) Conference, 384–93, Morgan-Kaufmann, San Mateo, CA.
30. Mead, C. and Mahowald, M. A. (1988) A silicon model of early visual processing. *Neural Networks*, **1**(1), 91–7.
31. Boser, B. E., Sackinger, E., Bromley, J. *et al.* (1991) An analog neural network processor with programmable topology. *IEEE J. Solid State Circ.*, **26**(12), 2017–25.
32. Mavor, J., Jack, M. and Denyer, P. (1983) *Introduction to MOS LSI Design*, Addison-Wesley, Reading, MA.
33. Holler, M., Tam, S., Castro, H. and Benson, R. (1989) *An Electrically Trainable Artificial Neural Network (ETANN) with 10240 'Floating Gate' Synapses*. International Joint Conference on Neural Networks (IJCNN89), 191–6, June.

34. Hopfield, J. (1990) The effectiveness of analog 'neural net' hardware. *Network*, 1(1), 27–40.
35. Mackie, W. S., Graf, H. P. and Denker, J. S. (1987) *Microelectronic Implementation of Connectionist Neural Network Models*. Neural Information Processing Systems (NIPS) Conference, 515–23.
36. Murray, A. F., Brownlow, M., Hamilton, A. et al. (1990) *Pulse-Firing Neural Chips for Hundreds of Neurons*. Neural Information Processing Systems (NIPS) Conference, 785–92, Morgan-Kaufmann, San Mateo, CA.
37. Thakoor, A. P., Lamb, J. L., Moopenn, A. and Lambe, J. (1986) Binary synaptic connections based on memory switching in a-Si:H, in *AIP Conference Proceedings 151, Neural Networks for Computing* (ed. John S. Denker), 426–431, American Institute of Physics.
38. Rose, M. J., Hajto, J., LeComber, P. G. et al. (1989) Amorphous silicon analogue memory devices. *J. Non-Cryst. Sol.*, **115**, 168–70.
39. Sage, J. P., Thompson, K. and Withers, R. S. (1986) *An Artificial Neural Network Integrated Circuit Based on MNOS/CCD Principles*. Proceedings of the AIP Conference on Neural Networks for Computing, Snowbird, UT, 381–5.
40. Vittoz, E., Oguey, H., Maher, M. A. et al. (1990) *Analog Storage of Adjustable Synaptic Weights*. Proceedings of the ITG/IEEE Workshop on Microelectronics for Neural Networks, Dortmund, Germany, 69–79, June.
41. Schmitt, O. H. (1937) An electrical theory of nerve impulse propagation. *Am. J. Physiol.*, **119**, 399–400.
42. Meador, J., Wu, A., Cole, C. et al. (1990) Programmable impulse neural circuits. *IEEE Trans. Neural Networks*, 2(1), 101–9.
43. Simmons, J. A. (1989) *Acoustic-Imaging Computations by Echolocating Bats: Unification of Diversely-Represented Stimulus Features into Whole Images*. Neural Information Processing Systems (NIPS) Conference, 2–9, Morgan-Kaufmann, San Mateo, CA.
44. Murray, A. F., Baxter, D. J., Butler, Z. F. et al. (1990) *Innovations in Pulse Stream Neural VLSI: Arithmetic and Communications*. IEEE Workshop on Microelectronics for Neural Networks, Dortmund, Germany, 8–15.
45. Del Corso, D., Gregoretti, F., Pellegrini, C. and Reyneri, L. M. (1990) *An Artificial Neural Network Based on Multiplexed Pulse Streams*. Proceedings of the ITG/IEEE Workshop on Microelectronics for Neural Networks, Dortmund, Germany, 28–39, June.
46. Siggelkow, A., Beltman, A. J., Nijhuis, J. A. G. and Spaanenburg, L. (1990) *Pulse-Density Modulated Neural Networks on a Semi-Custom Gate Forest*. Proceedings of the ITG/IEEE Workshop on Microelectronics for Neural Networks, Dortmund, Germany, 16–27, June.
47. El-Leithy, N., Zaghloul, M. and Newcomb, R. W. (1988) *Implementation of Pulse-Coded Neural Networks*. Proceedings of the 27th Conference on Decision and Control, 334–336.
48. Tomberg, J. (1989) *Fully Digital Neural Network Implementation Based on Pulse Density Modulation*. IEEE Custom Integrated Circuits Conference, San Diego, CA, 12.7.1–12.7.2, May.
49. Hirai, Y. (1989) *A Digital Neuro-chip with Unlimited Connectability for Large Scale Neural Networks*. Proceedings of the IJCNN, Washington, DC, 163–9.

REFERENCES

50. Daniell, P. M., Waller, W. A. J. and Bisset, D. A. (1989) *An Implementation of Fully Analogue Sum-of-Product Neural Models*. 1st IEE Conference on Artificial Neural Networks, 52–56.
51. Elias, J. G., Chu, H. H. and Meshreki, S. M. (1992) *A Neuromorphic Impulsive Circuit for Processing Dynamic Signals*. International Conference on Circuits and Systems, San Diego, CA, 2208–11.
52. DeYong, M. and Fields C. (1992) *Applications of Hybrid Analog-Digital Neural Networks in Signal Processing*. International Conference on Circuits and Systems, San Diego, CA, 2212–15.
53. Watola, D., Gembala, D. and Meador, J. *Competitive Learning in Asynchronous Pulse Coded Neural ICs*. International Conference on Circuits and Systems, San Diego, CA, 2216–19.
54. Lazzaro, J. (1992) *Low-Power Silicon Spiking Neurons and Axons*. International Conference on Circuits and Systems, San Diego, CA, 2220–3.
55. Moon, G. and Zaghloul, M. E. (1992) *CMOS Design of Pulse Coded Adaptive Neural Processing Elements Using Neural-Type Cells*. International Conference on Circuits and Systems, San Diego, CA, 2224–7.
56. de Savigny, M. and Newcomb, R. W. (1992) *Realization of Boolean Functions Using a Pulse Coded Neuron*. International Conference on Circuits and Systems, San Diego, CA, 2228–31.
57. Tomberg, J. and Kaski, K. (1992) *Feasibility of Synchronous Pulse-Density Modulation Arithmetic in IC Implementations of Artificial Neural Networks*. International Conference on Circuits and Systems, San Diego, CA, 2232–5.
58. Linares-Barranco, B., Rodriguez-Vazquez, A., Huertas, J. L. and Sanchez-Sinencio, E. (1992) *CMOS Analog Neural Network Systems Based on Oscillatory Neurons*. International Conference on Circuits and Systems, San Diego, CA, 2236–9.
59. Horowitz, P. and Hill, W. (1989) *The Art of Electronics*, Cambridge University Press, UK.
60. Stark, H. and Tuteur, F. B. (1979) *Electrical Communication – Theory and Systems*, Prentice-Hall, Englewood Cliffs, NJ.
61. Murray, A. F. and Smith, A. V. W. (1987) Asynchronous arithmetic for VLSI neural systems. *Electronics Lett.*, **23**(12), 642–3 June.
62. Murray, A. F. and Smith, A. V. W. (1987) *A Novel Computational and Signalling Method for VLSI Neural Networks*. European Solid State Circuits Conference, 19–22, VDE-Verlag, Berlin.
63. Murray, A. F. and Edwards, P. J. (1993) *Synaptic Weight Noise During MLP Training Enhances Fault Tolerance*. Neural Information Processing Systems (NIPS) Conference, 491–498, Morgan-Kaufmann, San Mateo, CA.
64. Murray, A. F. and Smith, A. V. W. (1988) Asynchronous VLSI neural networks using pulse stream arithmetic. *IEEE J. Solid-State Circ. & Syst.*, **23**(3), 688–97.
65. Hanson, J. E., Skelton, J. K. and Allstodt, D. J. (1989). *A Time-Multiplexed Switched-Capacitor Circuit for Neural Network Applications*. Proceedings of IEEE ISCAS, Portland, Oregon, 2177–80.
66. Massengill, L. W. (1990) *Multiplexed, Charge-Based Circuits for Analog Neural Systems*. Proceedings of the IJCNN-90, Washington, DC, II-88–II-91.
67. Hopfield, J. J. and Tank, D. W. (1985) Neural computation of decisions in optimisation problems. *Biol. Cybern.*, **52**, 141–52.

REFERENCES

68. Sivilotti, M. A., Emerling, M. R. and Mead, C. (1986) *VLSI Architectures for Implementation of Neural Networks*. Proceedings of the AIP Conference on Neural Networks for Computing, Snowbird, UT, 408–13.
69. Vittoz, E. (1989) Analog VLSI Implementations of Neural Networks, in *Journées d'Electronique 1989*, Presses Polytechniques Romandes, Lausanne.
70. Murray, A. F., Hamilton, A., Baxter, D. J. et al. (1992) Integrated pulse-stream neural networks – results, issues and pointers. *IEEE Trans. Neural Networks*, 385–93.
71. Schneider, C. R. and Card, H. C. (1992) *Analog CMOS Contrastive Hebbian Networks*. Applications of Artificial Neural Networks III SPIE Proceedings, 1709.
72. Murray, A. F. (1992) Multi-layer perceptron learning optimised for on-chip implementation – a noise-robust system. *Neural Computation*, **4**(3), 366–81.
73. Tarassenko, L., Marshall, G. F., Murray, A. F. and Gomez-Casteneda, F. (1992) Parallel analogue computation for real-time path planning, in *VLSI for Artificial Intelligence* (eds. J. G. Delgado-Frias and W. R. Moore), Plenum Press, New York, NY, 93–99.
74. Tarassenko, L. and Marshall, G. (1991) *Robot Path Planning using Resistive Grids*. 2nd IEE Conference on Artificial Neural Networks, 149–52.
75. Hutchinson, J., Koch, C., Luo, J., Mead, C. et al. (1988) Computing motion using analog and binary resistive networks. *IEEE Computer Mag.*, 52–63, April.
76. Prescott, T. and Mayhew, J. (1991) *Obstacle Avoidance through Reinforcement Learning*. Neural Information Processing Systems (NIPS) Conference, 523–30, Morgan-Kaufmann, San Mateo, CA.
77. Le Cun, Y., Boser, B., Denker, J. S. et al. (1990) *Handwritten Character Recognition with a Back-Propagation Network*. Neural Information Processing Systems (NIPS) Conference, Morgan-Kaufmann, San Mateo, CA.
78. Platt, J. (1991) A resource-allocating network for function interpolation. *Neural Computation*, **3**, 213–25.
79. Roberts, S. J. (1991) Analysis of the human sleep electroencephalogram using a self-organizing neural network. D.Phil thesis, University of Oxford.
80. Le Cun, Y., Boser, B., Denker, J. S. et al. (1989) Backpropagation applied to handwritten zip code recognition, *Neural Computation*, **1**, 541–51.
81. Renshaw, D., Denyer, P. B., Wang, G. and Lu, M. (1990) *ASIC image sensors*. Proceedings of the IEEE Symposium on Circuits and Systems, 3038–41.
82. Hollis, P. W., Harper, J. S. and Paulos, J. J. (1990) The effects of precision constraints in a back-propagation learning network, *Neural Computation*, **2**, 363–73.
83. Dolenko, B. K. and Card, H. C. (1993) *The Effects of Analog Hardware Properties on Backpropagation Networks with On-Chip Learning*, IEEE International Conference on Neural Networks, San Francisco, CA.
84. Shoemaker, P. A., Carlin, M. J. and Shimabukuro, R. L. (1991) Back propagation learning with trinary quantization of weight updates. *Neural Networks*, **4**, 231–41.
85. Alspector, J., Gannett, J. W., Haber, S. et al. (1991) A VLSI-efficient technique for generating multiple uncorrelated noise sources and its application to stochastic neural networks. *IEEE Trans. Circ. & Syst.*, **38**(1), 109–23.
86. Murray, A. F. (1989) Pulse arithmetic in VLSI neural networks. *IEEE Micro*, **9**(6), 64–74.

87. Jabri, M. and Flower, B. (1991) Weight perturbation: an optimal architecture and learning technique for analog VLSI feedforward and recurrent multilayer networks. *Neural Computation*, **3**, 546–65.
88. Hoehfeld, M. and Fahlman, S. E. (1992) Learning with limited numerical precision using the cascade-correlation algorithm. *IEEE Trans. Neural Networks*, **3**, 603–11.
89. Flower, B. and Jabri, M. (1993) Summed weight neuron perturbation: an O(N) improvement over weight perturbation. *Neural Information Processing Systems*, (NIPS) Conference, 212–219, Morgan-Kaufmann, San Mateo, CA.
90. Taylor, J. G. (1972) Spontaneous behaviour in neural networks. *J. Theor. Biol.*, **36**, 513–28.
91. Hanson, S. J. (1990) A stochastic version of the delta rule. *Physica D*, **42**, 265–72.
92. Sequin, C. H. and Clay, R. D. (1991) Fault tolerance in feed-forward artificial neural networks, in *Neural Networks: Concepts, Applications and Implementations*, **4**, pp. 111–141, Prentice-Hall, Englewood Cliffs, NJ.
93. Murray, A. F. and Edwards, P. J. (1993) Synaptic weight noise during MLP training: enhanced MLP performance and fault tolerance resulting from synaptic weight noise during training, *IEEE Trans. Neural Networks*. (In press.)

Index

Accuracy
  amorphous silicon device 35
  of computation, switched-capacitor 71
  in pulse modulation techniques 43, 44
Addition, pulse stream techniques 44–7
AGVs 96–102
  obstacle-detection/avoidance 97, 101–2
  path planning 98–9
Algorithm, see Back-propagation
Amorphous silicon device 34–7
  accuracy enhancing 35
Amplifier, dynamic 74
Analogue
  computation, advantages and disadvantages 40
  /digital combinations 24–5, 39–40
  hardware, back-propagation implementation 105–25
  multiplication, digital control 38
  multipliers 25
    advantages 13–14
    MOSFET 26
  neural network accelerator, see ANNA
ANNA 24–5
Autonomous Guided Vehicles, see AGVs
Available bandwidth 50

Back-propagation 4
  hardware implementation 105–20
  viability 121–5
Bandwidth, see Available bandwidth
Binary
  memory device 34

pulses, hybrid networks 39
Biological nervous system 38, 39
Bottom-up approach, hardware learning 116–17
Bulk terminal, see Substrate terminal, MOSFET

Cascadability, problems with 58, 59
CCD structure, MNOS device in 32
Channel utilization, pulse stream 50
Charge-balancing techniques 30
Cheque-reader 23–4
Chip
  CNAPS 19–21
  -in-the-loop technique 34, 114
  -to-chip communication 49, 51–3
  see also EPSILON
Chips
  process variation between 60
  test, pulse stream studies 75–6
Circuit design 55
  PFM neuron 66
CNAPS chips 19–21
Communication, pulse stream 49–51
Computation
  accuracy, switched-capacitor design 71
  digital, for and against 40
  per-pulse 71–2
Connected network of adaptive processors, see CNAPS chips
Content-addressable memory, see Memory
Current
  leakage in digital weight storage 30, 70–1
  pulse addition 47

INDEX

DACs, see Digital to analogue converters
Data corruption
  digital weight storage 30
Design tips, VLSI 92–3
Deterministic network, Hopfield 2–3
Digital
  /analogue combinations 24–5, 39–40
  to analogue converters, pulse stream studies 77
  computation, advantages and disadvantages 40
  hardware 13, 19
  weight storage 29
Distributed feedback 58
Drain
  source current, MOSFET 17–18, 22
  terminal, MOSFET 15–16
Dynamic
  amplifier 74
  weight storage 29–30

EEPROM
  analogue weight storage 30
  floating gate technology 33
Electrically Erasable Programmable Read Only Memory, see EEPROM
Electronic programmability, see Programmability
Electronically-Trainable Artificial Neural Network, see ETANN
Encoder–decoder tasks, hardware learning 108
EPSILON 55
  ANNA device 25
  neuron designs 88
  pulse stream case studies 85–92
  specification 90
Error
  correction 6
  function, noise injection 126, 132
  in hardware learning 117–25
  minimization 9–10
    using noise 112–15
    see also Back-propagation
  threshold 10
ES2, see European Silicon Structures
ETANN 27–8
  floating gate technology 34
  specification compared with EPSILON 90–1
European Silicon Structures (ES2) 55

Exclusive-OR
  pattern classification 6

Facial images, testing networks on 127–32
Fault tolerance, noise injection 129
Feedback
  distributed, see Distributed Feedback
  operational amplifier 61
Feedforward networks, see Networks, feedforward
Flat-band voltage 16
Floating gate technology 33–4
  disadvantages 34
  MNOS device 31
Forming process, amorphous silicon 34–7
Function nodes, RBF classifier 7

Gate
  source voltage 16–17, 22
    modulating potential in MOS transistor 30–1
  terminal, MOSFET 15–16
Generalization
  ability 129–32
  definition 5
Gilbert Multiplier 23

Handshaking, pulse stream communication 52
Hardware
  co-processors 102–3
  digital 13, 19
  learning 105–35
    test problem 117–19
    weight update 116, 119–21
Hopfield network 1–6
  deterministic model 2–3
  silicon 4
  stochastic model 1–2
hybrid approach
  AGVs 97–8
  see also Networks, hybrid

Imprecise multiplier, see Low-precision multiplier
Incremental learning 104
Input patterns, linearly separable 6
Inter-chip communication, see Chip-to-chip communication

K-means algorithm 7

Leakage, weight storage 30, 70–1
'Leaky' integrator 57, 58
Learning 104–35
  hardware 105
  incremental/sequential 104–5
  multi-layer perceptrons 105–6
  on-chip 104
  top-down approach 106–16
  virtual targets scheme 106, 107–16
  see also Training
Least-Mean Square rule 6–7
Least Significant Bit, *see* LSB
Linear region MOSFET 17, 18
LMS, *see* Least-Mean Square rule
Localization, AGVs 99–101
Low
  -level processing, AGVs 97
  -precision multiplier 27
LSB, hardware learning 109

Madaline Rule III 125
MDAC, ANNA 25
Mead, C., circuit and chip design 22, 23
Memory
  binary 34
  content-addressable
    Hopfield network 1–2, 4
Metal
  deposition of, amorphous silicon device 36–7
  nitride oxide silicon, *see* MNOS
  oxide silicon field effect transistors, *see* MOSFET
Minimum distance classifiers 94–5
MLP, *see* Multi-layer perceptrons
MNOS
  charge in nitride layer 31
  device in CCD structure 32
  networks 30–2
MOS transconductance multiplier, *see* Transconductance multiplier, MOS
MOSFET 15–28
  analogue multiplier 26
  behaving as a resistor 18
  drain
    source voltage 17–18, 22
    terminal 15–16
  gate
    source voltage 17

  terminal 15–16
  linear region 17, 18
  N-type/P-type 15, 16
  threshold voltage 16, 19
MOSIS digital process 61
Multi-layer perceptron 4, 8–10
  EPSILON 91
  pulse stream studies 71–2
  using with hardware learning 105–35
Multiplexing 50–3
  multi-layer networks 73
  loss of precision 52–3
Multiplication, pulse stream signals 47–9
  analogue 38
Multipliers
  analogue 13–14, 25, 26
  see also Gilbert Multiplier; Low-precision multiplier; Transconductance multiplier
Multi-wafer system 21

Navigation, robots 104
  software feasibility study 117–19
Negative terminals, *see* MOSFET, N-type/P-type
Nervous system, *see* Biological nervous system
Network
  accelerator, *see* ANNA
  Hopfield 1–4
    deterministic 2–3
    silicon 4
    stochastic 1–2
  minimum distance classifier 94–5
  optical character recognition 21, 25
  resistive grid 99
  resistively-connected 3
  simulation, adding noise 127–32
  virtual targets 115–16
Networks
  analogue-digital hybrids 39
  feedforward 8
  first generation 1–4
  hybrid 39–40
  multi-layer, *see* Multi-layer perceptrons
  neural VSLI 15–28
  noise immunity 22
  single-layer 5–7
Neuron
  designs, EPSILON 88
Nitride layer, MOS transistor 30

Noise
  adding to network: simulation 127–32
  effects on network 22
  in hardware/MLP learning experiments 108–10, 112–15, 125–33
  injection into error function 126, 132
    fault tolerance 129
  in pulse modulation techniques 42–3
  in training, conclusions 133–4
  see also Noisy synapses
Noisy synapses 125–33
  mathematical predictions 126–7
N-type MOSFET, *see* MOSFET, N-type/P-type

Obstacle-detection, AGVs 97, 101–2
OCR, *see* Optical character recognition
On-chip learning, *see* Learning, on-chip
Op-amp/resistor network 21–2
Optical character recognition network 21, 25
OR-based add function, pulse stream 45–7
Output pulse, pulse stream studies 82

PAM, *see* Pulse amplitude modulation
Parity tasks, hardware learning 108
Path planning, AGVs 98–9
Pattern classification 4–11
  EPSILON 91
  Exclusive-OR 6
  see also Input patterns
Perceptron
  convergence theorem 6
  definition 5
  learning rule, *see* Perceptron convergence theorem
  multi-layer 4, 8–10
Per-pulse computation 71–2
PFM, *see* Pulse frequency modulation
Phase modulation 42
Polysilicon layer, MOS device 31
Process variation
  between chips 60
  PWM neuron circuit 69
  voltage integrator 62
Programmability
  of amorphous silicon 34
  accuracy 36
  incremental approach 35
  pulse stream technique 69–71
  using MNOS technique 32
Programmable gain 66
P-type MOSFET, *see* MOSFET, N-type/P-type
Pulse
  amplitude
    modulation 41
    multiplication 48–9
  binary, hybrid networks 39
  frequency modulation 42
  neuron 64–6, 90
  height modulation 47
  modulation techniques 41–4
  regeneration 75
  stream 38–53
    addition 43–7
    asynchronous 51
    case studies 54–93
    channel utilization 50
    communication 49–51
    EPSILON 85–92
    handshaking 52
    history 38–40
    multiplexing 50–3
    multiplication 47–9
    neuron circuit 64–6
    synchronous 50
    testing 78–80
  width modulation 41
    neuron 67–9, 88
  width multiplication 47, 48
PWM, *see* Pulse width modulation

Radial Basis Function, *see* RBF classifiers
RBF classifiers 7
Ready to Receive, *see* RTR/RTT
Ready to Transmit, *see* RTR/RTT
Real-time processing 96
Resistive grid, AGVs 98–9
Retinas, silicon 24
Robot
  navigation 104
  see also AGVs
RTR/RTT 52

Sample-and-hold test, pulse stream studies 78
Saturated MOSFET 17
Sequential learning 104

Sigmoid
    function
        Hopfield network 3
        multi-layer networks 8
        pulse stream studies 75
        voltage integrator 62
    transfer function, per pulse computation 81
Silicon
    first generation network 4
    early work in 22
    retinas (Mead) 24
    why use it? 11–13
    see also Amorphous silicon device
Single-layer networks, see Networks, single-layer
Source
    terminal, MOSFET 15–16
    voltage, MOSFET 19
Speech 94–6
    see also Vowels
Speed
    of amorphous silicon devices 35
    of learning with weight perturbation 125
    in pulse modulation techniques 43–4, 67
SPICE simulations 56, 58
    pulse-stream studies 75
Stochastic
    Hopfield network 1–2
    learning systems, noise 115
Storage of analogue synaptic weights 29
Substrate terminal, MOSFET 15–16
Subthreshold circuits 22–4
Switched-capacitor
    design 69–71
    test chips, AGVs 99
Synapse, pulse stream studies 73–4
    design in EPSILON 87
Synaptic weight 1
    incremental learning 104
    incrementing/decrementing 116
    leakage 70–1
    linearity 70, 78
    perturbation, hardware learning 116–17, 118–21, 124–5
    precision, pulse stream studies 82–3
    sequential learning 104–5
    storage 29–37, 70–1
    update
        effects of noise 132–3
        hardware learning 116, 119–21
        probabilistic 124
        pulse stream studies 84

Test chips, see Chips, test
Threshold voltage 16
    problems 19
Time division multiplexing, see Multiplexing
Top-down approach to learning 106–16
Training
    AGVs 99
    MLP/EPSILON 91
    noise in 133–4
    pattern classification 5, 6
    perceptron learning rule 6
    time, injected noise levels 132–4
    see also ETANN; Learning; Unsupervised learning
Transconductance multiplier
    MOS 25
        gate source voltage 30–1
        use in synapse design 56–7
Transistor threshold, MOSFET, see Threshold voltage
Transversal filters 25
Tunnelling
    in floating gate technology 33
    MNOS device 31

Ultrasound, AGVs 99
Unsupervised learning 7

Variations between chips 60–1
VCO 57, 64
Virtual targets learning scheme 106, 107–16
VLSI
    applications 94–103
    design tips 92–3
    neural networks, see Networks, neural VLSI
Voltage
    controlled oscillator, see VCO
    integrator 61
Vowels
    classification of, EPSILON 91
    sound recognition of 111–12
    see also Speech

Wafer Scale Integration 21
Weak inversion 22
Weight, *see* Synaptic weight
Wire-guided systems (AGVs) 97

Words, recognition, *see* Speech; Vowels
WSI, *see* Wafer Scale Integration